Food Safety and Protection

Food Safety and Protection

Owen Clark

Larsen & Keller
www.larsen-keller.com

Food Safety and Protection
Owen Clark
ISBN: 978-1-64172-675-7 (Hardback)

Larsen & Keller
Published by Larsen and Keller Education,
5 Penn Plaza,
19th Floor,
New York, NY 10001, USA

Cataloging-in-Publication Data

Food safety and protection / Owen Clark.
p. cm.
Includes bibliographical references and index.
ISBN 978-1-64172-675-7
1. Food--Safety measures. 2. Food security. 3. Food. 4. Food supply. I. Clark, Owen.
TX531 .F66 2022
363.192--dc23

For more information regarding Larsen and Keller Education and its products, please visit the publisher's website www.larsen-keller.com

Table of Contents

Preface

The scientific discipline which deals with the handling, preparation and storage of food is known as food safety. The primary focus of this discipline is to prevent the occurrence and spread of food-borne illnesses. Food can be contaminated in several ways such as physical contamination, biological contamination and chemical contamination. Food safety also seeks to prevent harm to consumers. The primary focus areas for protecting consumers are safety between industry and the market, and then between the market and the consumer. Food safety considerations between industry and market are processes related to food hygiene, labeling of food and pesticide residues. The market to consumer considerations are safe delivery and preparation of food. This book is compiled in such a manner, that it will provide in-depth knowledge about the theory and practice of food safety. It aims to equip students and experts with the advanced topics and upcoming concepts in this area. Those in search of information to further their knowledge will be greatly assisted by this book.

Given below is the chapter wise description of the book:

Chapter 1- The practices which aim at preserving the quality of food and prevent it from contamination and food borne illness during preparation, handling and storage are referred to as food safety. This is an introductory chapter which will briefly introduce diverse aspects of food safety as well as their management systems.

Chapter 2- There are a number of technologies and methods which are used to ensure the safety of food. It includes hurdle technology, food irradiation, canning, freezing, pasteurization, biopreservation, etc. This chapter closely examines these different food safety methods to provide an extensive understanding of the subject.

Chapter 3- The food can be preserved by using many preservation techniques such as low-temperature preservation, thermal and non-thermal processing of food, controlling water activity, fermentation, food additive, chemical preservation, etc. This chapter has been carefully written to provide an easy understanding of these techniques associated with preservation of food.

Chapter 4- Food packaging is the process of enclosing food to prevent it from damage, contamination, spillage and spoilage. This chapter delves into various aspects of food packaging such as active packaging, intelligent packaging, packaging machines, symbols of packages, materials used in packaging, etc. to provide an in-depth understanding of the subject.

Chapter 5- Allergens are the naturally-occurring proteins in food which cause irrational immune responses. The presence of some unwanted chemicals and harmful microorganisms in the food causes food contamination, food-borne illnesses and food allergies. The topics elaborated in this chapter will help in gaining a better perspective about food allergens, contamination and illness.

At the end, I would like to thank all those who dedicated their time and efforts for the successful completion of this book. I also wish to convey my gratitude towards my friends and family who supported me at every step.

Owen Clark

1

Introduction to Food Safety

The practices which aim at preserving the quality of food and prevent it from contamination and food borne illness during preparation, handling and storage are referred to as food safety. This is an introductory chapter which will briefly introduce diverse aspects of food safety as well as their management systems.

Food Safety refers to handling, preparing and storing food in a way to best reduce the risk individuals becoming sick from foodborne illnesses.

Food safety is a global concern that covers a variety of different areas of everyday life.

The principles of food safety aim to prevent food from becoming contaminated and causing food poisoning. This is achieved through a variety of different avenues, some of which are:

- Properly cleaning and sanitising all surfaces, equipment and utensils.
- Maintaining a high level of personal hygiene, especially hand-washing.
- Storing, chilling and heating food correctly with regards to temperature, environment and equipment.
- Implementing effective pest control.
- Comprehending food allergies, food poisoning and food intolerance.

Food as defined in Food Safety & Standards Act is any substance, whether processed, partially processed or unprocessed, which is intended for human consumption and includes primary food to the extent defined in the Act, genetically modified or engineered food or food containing such ingredients, infant food, packaged drinking water, alcoholic drink, chewing gum, and any substance, including water used in the food during its manufacture, preparation or treatment but does not include any animal feed, live animals unless they are prepared or processed for placing on the market for human consumption, plants, prior to harvesting, drugs and medicinal products, cosmetics, narcotic or psychotropic substances: Provided that the Central Government may declare, by notification in the Official Gazette, any other article as food for the purposes of this Act having regards to its use, nature, substance or quality.

Primary food means an article of food, being a produce of agriculture or horticulture or animal husbandry and dairying or aquaculture in its natural form, resulting from the growing, raising, cultivation, picking, harvesting, collection or catching in the hands of a person other than a farmer or fisherman.

Safe Food

The general public might consider that 'safe food' means zero risk (no risk at all). Whereas the food manufacturer would consider 'what is an acceptable risk?' The opinion expressed is that zero risk is not feasible given the range of food products available, the complexity of the distribution chain and human nature. Unfortunately there is no public consensus on what constitutes an acceptable risk. A difficulty that arises in manufacturing 'safe' food is that the consumer is a mixed population with varying degrees of susceptibility and general life style. Food with high levels of preservatives to reduce microbial growth is undesirable by the consumer and perceived as 'over processed' with 'chemical additives' The consumer pressure is for greater varieties of fresh and minimally processed foods, natural preservatives with a guarantee of absolute safety.

Unsafe Food

An article of food whose nature, substance or quality is so affected as to render it injurious to health by:-

- The article itself, or its package thereof, which is composed, whether wholly or in part, of poisonous or deleterious substance;
- The article consisting, wholly or in part, of any filthy, putrid, rotten, decomposed or diseased animal substance or vegetable substance;
- Virtue of its unhygienic processing or the presence in that article of any harmful substance;

- The substitution of any inferior or cheaper substance whether wholly or in part;
- Addition of a substance directly or as an ingredient which is not permitted;
- The abstraction, wholly or in part, of any of its constituents;
- The article being so coloured, flavoured or coated, powdered or polished, as to damage or conceal the article or to make it appear better or of greater value than it really is;
- The presence of any colouring matter or preservatives other than that specified in respect thereof;
- The article having been infected or infested with worms, weevils, or insects;
- Virtue of its being prepared, packed or kept under insanitary conditions;
- Virtue of its being mis-branded or sub-standard or food containing extraneous matter;
- Virtue of containing pesticides and other contaminants in excess of quantities specified by regulations.

Food safety is being challenged nowadays by the global dimensions of food supply chains, the need for reduction of food waste and efficient use of natural resources such as clean water. Food safety deals with safeguarding the own national food supply chain from the introduction, growth or survival of hazardous microbial and chemical agents. But within a larger international context, borders are fading and surely this is the case for foodstuffs which are an important globally traded commodity. There is great divergence in the degree of organization, infrastructure, teaching capacity across countries and food protection (food quality, food preservation, food safety) needs to be tackled globally.

Health Risks Associated with Food

Food and water may be associated with certain hazards due to poor handling and storage conditions or sometimes due to certain inherent food constituents. Potential health hazards associated with food are Physical, Chemical, Biological and Radiological.

Physical Hazardous Agents in Food and Water

Physical hazards are hard foreign objects that can cause physical injury or illness to the consumer. Physical hazard may be inherent to the food or ingredient. For example chaff, straw and stalks in plant produce; bone fragment, feather and hair in meat products. They could be extraneous for example, metal, glass, stones, soil, dirt, jewellery parts, rodent hairs and excreta, human hair and nails, staple pins, iron filings, plastic parts, wood, paper, insects etc. Physical hazards are easily recognized by consumers and do not need much expert help to detect.

Potential health hazards associated with food	
Physical	Glass, stones, metal, woo d, bone, feathers, hairs, plastic, parts of insect pests, insulation material, rodent hair, excreta, metal pieces, staple pins, iron filings.
Chemical	Naturally occurring toxins (bacterial and fungal toxins, biotoxins), heavy metals, pesticides, veterinary drug residues, antibiotics, growth regulators, chemical residues, allerqens.
Biological	Microbiological (pathogenic bacteria, parasites, protozoa, viruses,)
Radio – nuclides	Natural or man made

Chemical Hazardous Agents in Food and Water

Chemical hazards are toxic substances that are produced naturally, added intentionally, or unintentionally. These include heavy metals, natural toxins, sanitizers, pesticides, antibiotics and drugs. The origin of chemical hazards is:

- Agronomical: These include the residues of pesticides, fertilizers, fungicides, antibiotics and growth hormones used on food crops.
- Natural food toxins: Chemicals ortoxins produced by fish, shellfish toxins, mushroom toxins.
- Toxins produced by infecting or ingested microbes: Toxins produced by algae or diatoms consumed by fish, mycotoxins produced by mould and bacterial growth.
- Food allergens: These are natural constituents, mainly proteins, of certain foods like egg, fish, milk, peanuts, sesame seeds, soybean, tree nuts and wheat gluten, and certain preservatives and additives like sulphite.
- Factory chemical residues: Chemicals used for cleaning and sanitizing food contact surfaces, pest control chemicals, lubricants, coatings, paints, refrigerants and water treatment chemicals.

Biological Hazardous Agents in Food and Water

Food borne microbial infections and intoxications include a wide spectrum of illnesses that are a growing global public health risk. These diseases are caused by ingestion of foodstuffs contaminated with microorganisms or their toxic metabolites. The contamination may occur at any stage during the process of harvesting, handling, transport, storage, processing, distribution, and consumption of food. Both acute and chronic clinical manifestations of food borne illnesses are known. The best known acute clinical presentation of food borne disease involves gastrointestinal symptoms. Some well-known ailments could be attributed to the following classes of parasites and pathogens.

Viruses

Faecal contamination of food via water, soil or food handlers is reported to be the main cause of viral food borne diseases. Viral food borne diseases include hepatitis A (symptoms include fever, weakness, nausea, and jaundice), Norwalk agent or Small Round Shaped virus (causes gastrointestinal ailments like nausea, vomiting, diarrhoea, pain, headache, and fever), Rotaviruses (cause diarrhoea in infants and young children).

Bacteria

Bacterial food borne diseases can be divided into two categories, food infections, and food poisoning, caused as a result of toxins produced by the microbes and it singestion. Common bacteria that cause food borne infections include, Salmonella typhi, S. paratyphi, Escherichia coli, Shigella dysentariae, Campylobacterjejuni, Vibrio cholerae, V.parahemolyticus, Listeria monocytogenes, Aeromonas hydrophila, and Halicobacter pylori. Among E. coli, the enterotoxic, and enterohemorrahagic forms are the most virulent. Food poisoning is caused by the human ingestion of toxins produced by Staphylococcus aureus, Clostridium botulinum, and Clostridium perfringens in contaminated food. Common rickettsia that are involved in foodborne infections include well known Coxiella burnetti, the cause of Q fever, Rickettsia prowazekii, the cause of typhus fever, and Rickettsia rickettsii, the cause of Rocky mountain spotted fever.

Pathogenic Escherichia Coli and Salmonella, an Emerging Issue in Fresh Produce

Escherichia coli belongs to the normal intestinal microbiota of humans and animals and the majority of them are not harmful. Certain *E. coli* however harbor virulence factors and can cause intestinal and extra-intestinal diseases. Shiga toxin-producing *Escherichia coli* are zoonotic bacteria, and thus often associated with foods of animal origin (especially beef and sheep meat). However, *E. coli* O157:H7 shed in cattle manure can survive for extended periods of time and intervention strategies to control this pathogen at the source are critical as fresh produce crops are often grown in proximity to animal raising operations. Ravva and Korn evaluated whether neem (*Azadirachta indica*), known for its antimicrobial and insecticidal properties, can be used to amend manure to control *E. coli* O157:H7.

Foodborne outbreaks from fruits and vegetables and border rejections of fresh produce due to non-compliance testing are on the rise, generate economic loss and food waste, and lead to a loss of trust and confidence in the safety of fresh produce. A few studies in this Special Issue were focused on safety of fresh produce combining observations of the local situation with sampling and testing of crops and the production environment (soil, water, contact surfaces) to identify risk factors for contamination with enteric bacterial pathogens. Investigated the occurrence of generic *E. coli*, *Campylobacter*, *Salmonella* and pathogenic Shiga-toxin producing *E. coli* (STEC) on strawberries and the strawberries' production environment. Assembled the results from a multi-country study and found that generic *E. coli* was a suitable index organism for *Salmonella* and STEC when sampling leafy greens and strawberries at primary production and production environment, but to a lesser extent for *Campylobacter*.

Protozoa and Parasites

Several parasites are known to cause foodborne and waterborne illnesses. These organisms live and reproduce within the tissues and organs of animal and human hosts, and are transmitted to humans through the contaminated food and water. Parasites are of different types and range in size, from single- celled microscopic organisms (protozoa), to large multi-cellular worms (helminthes) that may be seen without a microscope. Amoebic dysentery is the most common ailment caused by Entamoeba histolytica. The symptoms include diarrhea, fever, chills, and liver abscess. Giardiasis caused by Giardia causes diarrhea with green stools. These diseases are caused by raw or mishandled food or contaminated water.

Toxoplasmosis caused by Toxoplasma gondii causes mononucleosis and death. The disease is caused by consumption of raw undercooked meat. Trichinella spiralis, cause of trichinellosis (also known as trichinosis), is an intestinal roundworm whose larvae may migrate from the digestive tract and form cysts in various muscles of the body. Infections occur worldwide, but are most prevalent in regions where pork or wild game is consumed raw or undercooked. Taenia saginata and Taenia solium are parasitic helminthes. Taeniasisisthe intestinal infection caused by adult-stage tapeworms.

Yeasts, Moulds and Mycotoxins

Their prolonged presence, particularly of the Mycotoxins producing moulds may lead to formation of Mycotoxins. Mycotoxins can appear in the food chain as a result of pre-harvest fungal infection of crops, or during postharvest storage, if conditions are conducive for fungal growth. Once formed Mycotoxins largely are difficult to eliminate. So, they tend to remain in the food chain, and in meat and dairy products. Even drastic thermal treatments, such as cooking do not destroy some Mycotoxins. Therefore, prevention of mould growth in pre-harvest and postharvest stored crops is the best strategy to eliminate Mycotoxins in food chain.

Radiological Hazardous Agents in Food and Water

Natural: All food and drinking water may contain some level of natural radioactivity due to the presence of naturally occurring radionuclides like potassium-40, and radium-226.

Manmade: Foods and water may get contaminated with radionuclides due to their release in environment from nuclear establishments and in large quantities from nuclear accidents. In case of nuclear accidents, reactor produced radionuclides such as Iodine-131, Cesium-137, Strontium-90, may find entry in food chain.

Contamination of the food or feed with radionuclides may impact human or animal health and also become barrier to trade. Contamination of foods with radionuclide shall be as low as reasonably achievable.

Hazardous substances can enter at any point along the food chain, i.e. from farm where food is produced to the consumer.

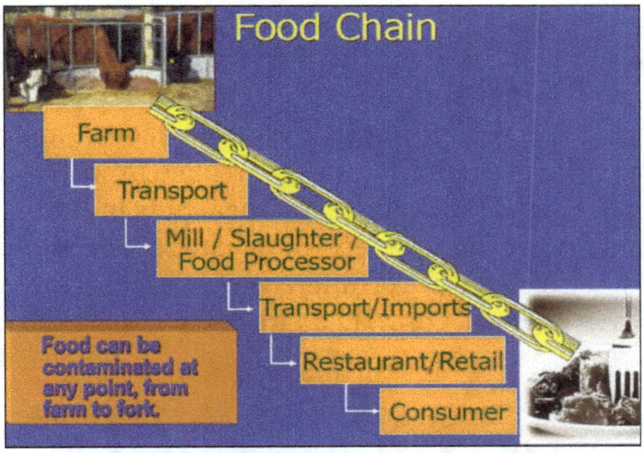

Food chain and points of entry of hazardous substances.

Rationale

All ports of entry should have surveillance and monitoring program for providing safe food for international travellers including food in transit. Food business operators should follow an established Food Safety Management System, use a HACCP Protocol to ensure production of safe food. Further, It is essential that food service personnel protect the health of consumers from food borne illnesses by practicing clean work habits and following acceptable sanitary procedures, thereby, creating a sanitary environment.

Water Treatment Technologies

Leafy greens remain one of the most relevant crops in fresh produce, with an increasing production of bagged salads. However, increased vigilance, in particular related to the selection of water sources and use of water treatment, may help to take safety of fresh produce to the next level. Appropriate irrigation management practices are needed to guarantee the sustainability of the environment and the quality of the leafy greens. The study by Allende and Monaghan provides an overview of the main problems in the production of leafy vegetables associated with irrigation water, including microbial risk and difficulties in water monitoring, compliance with evolving regulations and quality standards, and summarizes the current alternatives available for growers to reduce microbial risks. In addition the Special Issue also includes a dedicated study presenting the data on fecal indicators and selected bacterial pathogens to assess the level of fecal contamination of a Norwegian river used for irrigation in an area which has a high production level of various types of food commodities.

The importance of water quality for the rinse step of leafy greens at harvest and also the washing step in the production of fresh-cut produce was identified as a potential pathway for dispersion of spoilage bacteria, fecal indicator bacteria such as E. coli or introduction of pathogens via cross-contamination. This was observed, in particular, if no sanitizing agents were used to keep the water clean in the washing tank. Effectiveness of water disinfection needs to be validated case by case, and its appropriate use monitored. The review by examines the efficacy of process wash water disinfectants during produce processing. The conditions for use of chlorine and a range of alternatives such as chlorine dioxide, peroxyacetic acid, ozone in wash water disinfection to prevent cross-contamination were studied. Another study in this Special Issue focused on decontamination of fresh-cut vegetables. evaluated the effect of combined decontamination methods based on the use of different sanitizers and the application of pressure on the inactivation of E. coli O157:H7 on fresh-cut lettuce and carrots.

Food Safety Management Systems

A food safety management system (FSMS) is not only a legal requirement, but a helpful tool to ensure safe practices are followed within your business. A FSMS is a systematic approach to controlling food safety hazards within a food business in order to ensure that food is safe to eat. An effective food safety management system that is based on seven principles of HACCP can help businesses to identify and control hazards before they threaten the safety of your food and customers.

HACCP

The Hazard Analysis and Critical Control Points (HACCP) system is a logical, scientific approach to controlling hazards in meat production. HACCP is a preventive system assuring the safe production of food products. The application of HACCP is based on technical and scientific principles that assure food safety. An ideal application would include all processes from the farm to the table. The principle of HACCP can be applied to production, meat slaughter and processing, shipping and distribution, food service and in home preparation.

HACCP is a systematic preventative system that uses common sense application of scientific principles. The most important aspect of HACCP is that it is a preventative system rather than an inspection system of controlling food safety hazards. Prevention of hazards cannot be accomplished by end product inspection, so controlling the production process with HACCP offers the best approach. The application of HACCP is systematic because structured hazard analysis and implementation are provided. The process is common sense in that each processor understands their operation and is best able to assess controlling the process. HACCP is also science-based and so the controls that are placed in the process should be based on scientific information.

The HACCP system has two major components. The HA of HACCP represents the logic in the hazard analysis which identifies the where and how of hazards. The CCP of HACCP represents the critical control points that provide the control of the process and the proof of the control. The end objective of HACCP is to make the product as safe as possible and to be able to prove that the product was processed as safe as possible. This does not mean that HACCP provides 100% assurance of food safety to consumers, but does mean that a meat processing company is doing the best job possible for safe food production.

The assurance of safety comes from the process of identifying the hazards, establishing controls for the identified hazards, monitoring the controls and periodically verifying that the system works.

Hazards

HACCP focuses on three types of hazards; biological hazards, chemical hazards, and physical hazards. Biological hazards are the type of hazards that receive the most attention in the HACCP system and also present the greatest risk of severity and occurance. Biological hazards include hazards from pathogens such as bacteria, viruses, yeasts and molds. Bacteria that receive the greatest attention in the United States include *E. coli* 0157:H7, *Listeria monocytogenes*, *Salmonella*, *Staphylococus auerus*, and *Campylobacter*. Chemical hazards in meat products could result from mis-use of antibiotics in production, contamination with sanitizers or cleaning agents, or environmental contamination from hydrolic fluids. Physical hazards are probably the most recognized by consumers as they usually find this hazard. Glass, metal, and plastic are physical hazards that can occur in meat products.

Developing a HACCP Plan

To start a HACCP system, a company must first write a HACCP plan. Companies may use generic models as resources for developing a plant specific plan, however, the most useful and successful HACCP plans need to be developed from the very beginning from the plant that will use and implement the plan. To develop a HACCP plan, a team of individuals from within the company, with some assistance from outside experts, conducts five preliminary steps and applies the seven HACCP principles.

The format of HACCP plans will vary. In many cases the plans will be product and process specific. However, some plans may use a unit operations approach.

Generic HACCP plans can serve as useful guides in the development of process and product HACCP plans; however, it is essential that the unique conditions within each facility be considered during the development of all components of the HACCP plan.

In the development of a HACCP plan, five preliminary tasks need to be accomplished before the application of the HACCP principles to a specific product and process. The five preliminary tasks are given in figure.

Preliminary Tasks in the Development of the HACCP Plan.

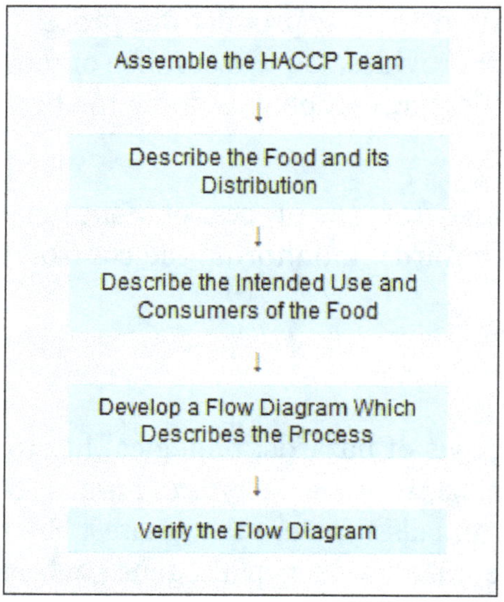

Assemble the HACCP Team

The first task in developing a HACCP plan is to assemble a HACCP team consisting of individuals who have specific knowledge and expertise appropriate to the product and process. It is the team's responsibility to develop the HACCP plan. The team should be multi disciplinary and include individuals from areas such as engineering, production, sanitation, quality assurance, and food microbiology. The team should also include local personnel who are involved in the operation as they are more familiar with the variability and limitations of the operation. In addition, this fosters a sense of ownership among those who must implement the plan. The HACCP team may need assistance from outside experts who are knowledgeable in the potential biological, chemical and physical hazards associated with the product and the process. However, a plan which is developed totally by outside sources may be erroneous, incomplete, and lacking in support at the local level.

Due to the technical nature of the information required for hazard analysis, it is recommended that experts who are knowledgeable in the food process should either participate in or verify the completeness of the hazard analysis and the HACCP plan.

Such individuals should have the knowledge and experience to correctly: (a) conduct a hazard analysis; (b) identify potential hazards; (c) identify hazards which must be controlled; (d) recommend controls, critical limits, and procedures for monitoring and verification; (e) recommend appropriate corrective actions when a deviation occurs; (f) recommend research related to the HACCP plan if important information is not known; and (g) validate the HACCP plan.

Describe the Food and its Distribution

The HACCP team first describes the food. This consists of a general description of the food, ingredients, and processing methods. The method of distribution should be described along with information on whether the food is to be distributed frozen, refrigerated, or at ambient temperature.

Describe the Intended use and Consumers of the Food

Describe the normal expected use of the food. The intended consumers may be the general public or a particular segment of the population (e.g., infants, immunocompromised individuals, the elderly, etc.).

Develop a Flow Diagram which Describes the Process

The purpose of a flow diagram is to provide a clear, simple outline of the steps involved in the process. The scope of the flow diagram must cover all the steps in the process which are directly under the control of the establishment. In addition, the flow diagram can include steps in the food chain which are before and after the processing that occurs in the establishment. The flow diagram need not be as complex as engineering drawings. A block type flow diagram is sufficiently descriptive. Also, a simple schematic of the facility is often useful in understanding and evaluating product and process flow.

Verify the Flow Diagram

The HACCP team should perform an on-site review of the operation to verify the accuracy and completeness of the flow diagram. Modifications should be made to the flow diagram as necessary and documented.

After these five preliminary tasks have been completed, the seven principles of HACCP are applied.

The seven HACCP principles are the most important steps in writing a HACCP plan. The first two steps provide the foundation for the HACCP plan. These two steps are essential since application of the other HACCP principles depend on the results of the hazard analysis. The remaining five steps are the application steps of the HACCP plan and provide the structure for conducting the workings of the HACCP plan int he processing plant.

HACCP Principles

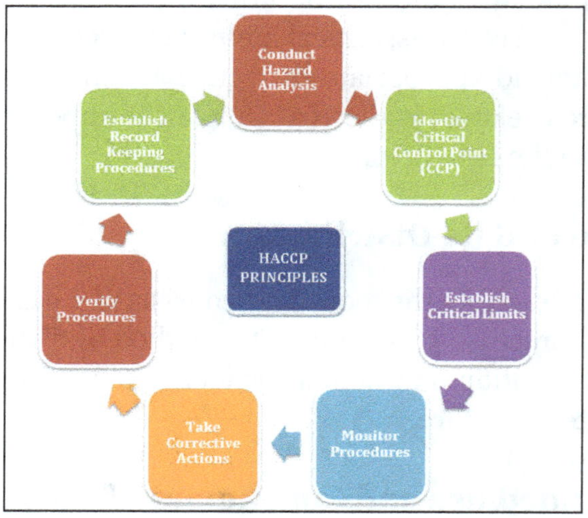

Conduct a Hazard Analysis (Principle 1)

After addressing the preliminary tasks discussed above, the HACCP team conducts a hazard analysis and identifies appropriate control measures. The purpose of the hazard analysis is to develop a list of hazards which are of such significance that they are reasonably likely to cause injury or illness if not effectively controlled. Hazards that are not reasonably likely to occur would not require further consideration within a HACCP plan. It is important to consider in the hazard analysis the ingredients and raw materials, each step in the process, product storage and distribution, and final preparation and use by the consumer. When conducting a hazard analysis, safety concerns must be differentiated from quality concerns. A hazard is defined as a biological, chemical or physical agent that is reasonably likely to cause illness or injury in the absence of its control. Thus, the word hazard as used in this document is limited to safety.

A thorough hazard analysis is the key to preparing an effective HACCP plan. If the hazard analysis is not done correctly and the hazards warranting control within the HACCP system are not identified, the plan will not be effective regardless of how well it is followed.

The hazard analysis and identification of associated control measures accomplish three objectives: Those hazards and associated control measures are identified. The analysis may identify needed modifications to a process or product so that product safety is further assured or improved. The analysis provides a basis for determining CCPs in Principle 2.

The process of conducting a hazard analysis involves two stages. The first, hazard identification, can be regarded as a brain storming session. During this stage, the HACCP team reviews the ingredients used in the product, the activities conducted at each step in the process and the equipment used, the final product and its method of storage and distribution, and the intended use and consumers of the product. Based on this review, the team develops a list of potential biological, chemical or physical hazards which may be introduced, increased, or controlled at each step in the production process. Hazard identification focuses on developing a list of potential hazards associated with each process step under direct control of the food operation. A knowledge of any adverse health-related events historically associated with the product will be of value in this exercise.

After the list of potential hazards is assembled, stage two, the hazard evaluation, is conducted. In stage two of the hazard analysis, the HACCP team decides which potential hazards must be addressed in the HACCP plan. During this stage, each potential hazard is evaluated based on the severity of the potential hazard and its likely occurrence. Severity is the seriousness of the consequences of exposure to the hazard. Considerations of severity (e.g., impact of sequelae, and magnitude and duration of illness or injury) can be helpful in understanding the public health impact of the hazard. When conducting the hazard evaluation, it is helpful to consider the likelihood of exposure and severity of the potential consequences if the hazard is not properly controlled.

In addition, consideration should be given to the effects of short term as well as long term exposure to the potential hazard. Such considerations do not include common dietary choices which lie outside of HACCP. During the evaluation of each potential hazard, the food, its method of preparation, transportation, storage and persons likely to consume the product should be considered to determine how each of these factors may influence the likely occurrence and severity of the hazard being controlled. The team must consider the influence of likely procedures for food preparation and storage and whether the intended consumers are susceptible to a potential hazard. However, there may be differences of opinion, even among experts, as to the likely occurrence and severity of a hazard. The HACCP team may have to rely upon the opinion of experts who assist in the development of the HACCP plan.

Hazards identified in one operation or facility may not be significant in another operation producing the same or a similar product. For example, due to differences in equipment and an effective maintenance program, the probability of metal contamination may be significant in one facility but not in another. A summary of the HACCP team deliberations and the rationale developed during the hazard analysis should be kept for future reference. This information will be useful during future reviews and updates of the hazard analysis and the HACCP plan.

Upon completion of the hazard analysis, the hazards associated with each step in the production of the food should be listed along with any measure(s) that are used to control the hazard(s). The term control measure is used because not all hazards can be prevented, but virtually all can be controlled. More than one control measure may be required for a specific hazard. On the other hand, more than one hazard may be addressed by a specific control measure (e.g. pasteurization of milk).

For example, if a HACCP team were to conduct a hazard analysis for the production of frozen cooked beef patties, enteric pathogens (e.g., *Salmonella* and verotoxin-producing *Escherichia coli*) in the raw meat would be identified as hazards. Cooking is a control measure which can be used to eliminate these hazards. The following is an excerpt from a hazard analysis summary table for this product.

Step	Potential Hazard(s)	Justification	Hazard to be addressed in plan? Y/N	Control Measure(s)
Cooking	Enteric pathogens: e.g., Salmonella, verotoxigenic-E. coli	Enteric pathogens have been associated with outbreaks of foodborne illness from undercooked ground beef	Y	Cooking

The hazard analysis summary could be presented in several different ways. One format is a table such as the one given above. Another could be a narrative summary of the HACCP team's hazard analysis considerations and a summary table listing only the hazards and associated control measures.

Determine Critical Control Points (CCPs) (Principle 2)

A critical control point is defined as a step at which control can be applied and is essential to prevent or eliminate a food safety hazard or reduce it to an acceptable level. The potential hazards that are reasonably likely to cause illness or injury in the absence of their control must be addressed in determining CCPs.

Complete and accurate identification of CCPs is fundamental to controlling food safety hazards. The information developed during the hazard analysis is essential for the HACCP team in identifying which steps in the process are CCPs. One strategy to facilitate the identification of each CCP is the use of a CCP decision tree. Although application of the CCP decision tree can be useful in determining if a particular step is a CCP for a previously identified hazard, it is merely a tool and not a mandatory element of HACCP. A CCP decision tree is not a substitute for expert knowledge.

Critical control points are located at any step where hazards can be either prevented, eliminated, or reduced to acceptable levels. Examples of CCPs may include: thermal processing, chilling, testing ingredients for chemical residues, product formulation control, and testing product for metal contaminants. CCPs must be carefully developed and documented. In addition, they must be used only for purposes of product safety. For example, a specified heat process, at a given time and temperature designed to destroy a specific microbiological pathogen, could be a CCP. Likewise, refrigeration of a precooked food to prevent hazardous microorganisms from multiplying, or the adjustment of a food to a pH necessary to prevent toxin formation could also be CCPs. Different facilities preparing similar food items can differ in the hazards identified and the steps which are CCPs. This can be due to differences in each facility's layout, equipment, selection of ingredients, processes employed, etc.

Establish Critical Limits (Principle 3)

A critical limit is a maximum and minimum value to which a biological, chemical or physical parameter must be controlled at a CCP to prevent, eliminate or reduce to an acceptable level the occurrence of a food safety hazard. A critical limit is used to distinguish between safe and unsafe operating conditions at a CCP. Critical limits should not be confused with operational limits which are established for reasons other than food safety.

Each CCP will have one or more control measures to assure that the identified hazards are prevented, eliminated or reduced to acceptable levels. Each control measure has one or more associated critical limits. Critical limits may be based upon factors such as: temperature, time, physical dimensions, humidity, moisture level, water activity (a_w), pH, titratable acidity, salt concentration, available chlorine, viscosity, preservatives, or sensory information such as aroma and visual appearance. Critical limits must be scientifically based. For each CCP, there is at least one criterion for food safety that is to be met. An example of a criterion is a specific lethality of a cooking process such as a 5D reduction in *Salmonella*. The critical limits and criteria for food safety may be derived from sources such as regulatory standards and guidelines, literature surveys, experimental results, and experts.

An example is the cooking of beef patties. The process should be designed to ensure the production of a safe product. The hazard analysis for cooked meat patties identified enteric pathogens (e.g., verotoxigenic *E. coli* such as *E. coli* O157:H7, and salmonellae) as significant biological hazards. Furthermore, cooking is the step in the process at which control can be applied to reduce the enteric pathogens to an acceptable level. To ensure that an acceptable level is consistently achieved, accurate information is needed on the probable number of the pathogens in the raw patties, their heat resistance, the factors that influence the heating of the patties, and the area of the patty which heats the slowest. Collectively, this information forms the scientific basis for the critical limits that are established. Some of the factors that may affect the thermal destruction of enteric pathogens are listed in the following table.

In this example, the HACCP team concluded that a thermal process equivalent to 155 °F for 16 seconds would be necessary to assure the safety of this product. To ensure that this time and temperature are attained, the HACCP team for one facility determined that it would be necessary to establish critical limits for the oven temperature and humidity, belt speed (time in oven), patty thickness and composition (e.g., all beef, beef and other ingredients). Control of these factors enables the facility to produce a wide variety of cooked patties, all of which will be processed to a minimum internal temperature of 155 °F for 16 seconds. In another facility, the HACCP team may conclude that the best approach is to use the internal patty temperature of 155 °F and hold for 16 seconds as critical limits. In this second facility the internal temperature and hold time of the patties are monitored at a frequency to ensure that the critical limits are constantly met as they exit the oven. The example given below applies to the first facility.

Process Step	CCP	Critical Limits
Cooking	YES	Oven temperature:____ °F Time; rate of heating and cooling (belt speed in ft/min): _____ft/min Patty thickness: _____in. Patty composition: e.g. all beef Oven humidity: _____% RH

Establish Monitoring Procedures (Principle 4)

Monitoring is a planned sequence of observations or measurements to assess whether a CCP is under control and to produce an accurate record for future use in verification. Monitoring serves three main purposes. First, monitoring is essential to food safety management in that it facilitates tracking of the operation. If monitoring indicates that there is a trend towards loss of control, then action can be taken to bring the process back into control before a deviation from a critical limit occurs. Second, monitoring is used to determine when there is loss of control and a deviation occurs at a CCP, i.e., exceeding or not meeting a critical limit. When a deviation occurs, an appropriate corrective action must be taken. Third, it provides written documentation for use in verification.

An unsafe food may result if a process is not properly controlled and a deviation occurs. Because of the potentially serious consequences of a critical limit deviation, monitoring procedures must be effective. Ideally, monitoring should be continuous, which is possible with many types of physical and chemical methods. For example, the temperature and time for the scheduled thermal process of low-acid canned foods is recorded continuously on temperature recording charts. If the temperature falls below the scheduled temperature or the time is insufficient, as recorded on the chart, the product from the retort is retained and the disposition determined as in Principle 5. Likewise, pH measurement may be performed continually in fluids or by testing each batch before processing. There are many ways to monitor critical limits on a continuous or batch basis and record the data on charts. Continuous monitoring is always preferred when feasible. Monitoring equipment must be carefully calibrated for accuracy.

Assignment of the responsibility for monitoring is an important consideration for each CCP. Specific assignments will depend on the number of CCPs and control measures and the complexity of monitoring. Personnel who monitor CCPs are often associated with production (e.g., line supervisors, selected line workers and maintenance personnel) and, as required, quality control personnel. Those individuals must be trained in the monitoring technique for which they are responsible, fully understand the purpose and importance of monitoring, be unbiased in monitoring and reporting, and accurately report the results of monitoring. In addition, employees should be trained in procedures to follow when there is a trend towards loss of control so that adjustments can be made in a timely manner to assure that the process remains under control. The person responsible for monitoring must also immediately report a process or product that does not meet critical limits.

All records and documents associated with CCP monitoring should be dated and signed or initialed by the person doing the monitoring.

When it is not possible to monitor a CCP on a continuous basis, it is necessary to establish a monitoring frequency and procedure that will be reliable enough to indicate that the CCP is under control. Statistically designed data collection or sampling systems lend themselves to this purpose.

Most monitoring procedures need to be rapid because they relate to on-line, "real-time" processes and there will not be time for lengthy analytical testing. Examples of monitoring activities include: visual observations and measurement of temperature, time, pH, and moisture level.

Microbiological tests are seldom effective for monitoring due to their time-consuming nature and problems with assuring detection of contaminants. Physical and chemical measurements are often preferred because they are rapid and usually more effective for assuring control of microbiological hazards. For example, the safety of pasteurized milk is based upon measurements of time and temperature of heating rather than testing the heated milk to assure the absence of surviving pathogens.

With certain foods, processes, ingredients, or imports, there may be no alternative to microbiological testing. However, it is important to recognize that a sampling protocol that is adequate to reliably detect low levels of pathogens is seldom possible because of the large number of samples needed. This sampling limitation could result in a false sense of security by those who use an inadequate sampling protocol. In addition, there are technical limitations in many laboratory procedures for detecting and quantitating pathogens and their toxins.

Establish Corrective Actions (Principle 5)

The HACCP system for food safety management is designed to identify health hazards and to establish strategies to prevent, eliminate, or reduce their occurrence. However, ideal circumstances do not always prevail and deviations from established processes may occur. An important purpose of corrective actions is to prevent foods which may be hazardous from reaching consumers. Where there is a deviation from established critical limits, corrective actions are necessary. Therefore, corrective actions should include the following elements: (a) determine and correct the cause of non-compliance; (b) determine the disposition of non-compliant product and (c) record the corrective actions that have been taken. Specific corrective actions should be developed in advance for each CCP and included in the HACCP plan. As a minimum, the HACCP plan should specify what is done when a deviation occurs, who is responsible for implementing the corrective actions, and that a record will be developed and maintained of the actions taken. Individuals who have a thorough understanding of the process, product and HACCP plan should be assigned the responsibility for oversight of corrective actions. As appropriate, experts may be consulted to review the information available and to assist in determining disposition of non-compliant product.

Establish Verification Procedures (Principle 6)

Verification is defined as those activities, other than monitoring, that determine the validity of the HACCP plan and that the system is operating according to the plan. The NAS pointed out that the major infusion of science in a HACCP system centers on proper identification of the hazards, critical control points, critical limits, and instituting proper verification procedures. These processes should take place during the development and implementation of the HACCP plans and maintenance of the HACCP system. An example of a verification schedule is given in table.

One aspect of verification is evaluating whether the facility's HACCP system is functioning according to the HACCP plan. An effective HACCP system requires little end-product testing, since sufficient validated safeguards are built in early in the process. Therefore, rather than relying on end-product testing, firms should rely on frequent reviews of their HACCP plan, verification that the HACCP plan is being correctly followed, and review of CCP monitoring and corrective action records.

Another important aspect of verification is the initial validation of the HACCP plan to determine that the plan is scientifically and technically sound, that all hazards have been identified and that if the HACCP plan is properly implemented these hazards will be effectively controlled. Information needed to validate the HACCP plan often include (1) expert advice and scientific studies and (2) in-plant observations, measurements, and evaluations. For example, validation of the cooking process for beef patties should include the scientific justification of the heating times and temperatures needed to obtain an appropriate destruction of pathogenic microorganisms (i.e., enteric pathogens) and studies to confirm that the conditions of cooking will deliver the required time and temperature to each beef patty.

Subsequent validations are performed and documented by a HACCP team or an independent expert as needed. For example, validations are conducted when there is an unexplained system failure; a significant product, process or packaging change occurs; or new hazards are recognized.

In addition, a periodic comprehensive verification of the HACCP system should be conducted by an unbiased, independent authority. Such authorities can be internal or external to the food operation. This should include a technical evaluation of the hazard analysis and each element of the HACCP plan as well as on-site review of all flow diagrams and appropriate records from operation of the plan. A comprehensive verification is independent of other verification procedures and must be performed to ensure that the HACCP plan is resulting in the control of the hazards. If the results of the comprehensive verification identifies deficiencies, the HACCP team modifies the HACCP plan as necessary.

Verification activities are carried out by individuals within a company, third party experts, and regulatory agencies. It is important that individuals doing verification have appropriate technical expertise to perform this function.

Table: Example of a company established HACCP verification schedule.

Activity	Frequency	Responsibility	Reviewer
Verification Activities Scheduling.	Yearly or Upon HACCP System Change.	HACCP Coordinator.	Plant Manager.
Initial Validation of HACCP Plan.	Prior to and During Initial Implementation of Plan.	Independent Experts.	HACCP Team.
Subsequent validation of HACCP Plan.	When Critical Limits Changed, Significant Changes in Process, Equipment Changed, After System Failure, etc.	Independent Experts.	HACCP Team.
Verification of CCP Monitoring as Described in the Plan (e.g., monitoring of patty cooking temperature).	According to HACCP Plan (e.g., once per shift).	According to HACCP Plan (e.g., Line Supervisor).	According to HACCP Plan (e.g., Quality Control).
Review of Monitoring, Corrective Action Records to Show Compliance with the Plan.	Monthly.	Quality Assurance.	HACCP Team.
Comprehensive HACCP System Verification.	Yearly.	Independent Experts.	Plant Manager.
Done by others than the team writing and implementing the plan. May require additional technical expertise as well as laboratory and plant test studies.			

Establish Record-keeping and Documentation Procedures (Principle 7)

Generally, the records maintained for the HACCP System should include the following:

- A summary of the hazard analysis, including the rationale for determining hazards and control measures.
- The HACCP Plan:
  - Listing of the HACCP team and assigned responsibilities.
  - Description of the food, its distribution, intended use, and consumer.
  - Verified flow diagram.

HACCP plan summary table that includes information for steps in the process that are CCPs:

- The hazard(s) of concern.
- Critical limits.
- Monitoring.
- Corrective actions.
- Verification procedures and schedule.
- Record-keeping procedures.

A brief summary of position responsible for performing the activity and the procedures and frequency should be provided.

The following is an example of a HACCP plan summary table:

CCP	Hazards	Critical limit(s)	Monitoring	Corrective Actions	Verification	Records

- Support documentation such as validation records.
- Records that are generated during the operation of the plan.

Implementation and Maintenance of the Haccp Plan

The successful implementation of a HACCP plan is facilitated by commitment from top management. The next step is to establish a plan that describes the individuals responsible for developing, implementing and maintaining the HACCP system. Initially, the HACCP coordinator and team are selected and trained as necessary. The team is then responsible for developing the initial plan and coordinating its implementation. Product teams can be appointed to develop HACCP plans for specific products. An important aspect in developing these teams is to assure that they have appropriate training. The workers who will be responsible for monitoring need to be adequately trained. Upon completion of the HACCP plan, operator procedures, forms and procedures for monitoring and corrective action are developed. Often it is a good idea to develop a timeline for the activities involved in the initial implementation of the HACCP plan. Implementation of the HACCP system involves the continual application of the monitoring, record-keeping, corrective action procedures and other activities as described in the HACCP plan.

Maintaining an effective HACCP system depends largely on regularly scheduled verification activities. The HACCP plan should be updated and revised as needed. An important aspect of maintaining the HACCP system is to assure that all individuals involved are properly trained so they understand their role and can effectively fulfill their responsibilities.

References

- What-is-food-safety, resources: foodsafety.com.au, Retrieved 4 May 2018
- Food-water-safety: ihrpoe.co.in, Retrieved 16 June 2018
- Food-safety-for-businesses, food-safety-589: bromley.gov.uk, Retrieved 09 August 2018
- Introduction-haccp-training: food.unl.edu, Retrieved 27 March 2018
- Haccp-principles-application-guidelines, hazard-analysis-critical-control-point-haccp: da.gov, Retrieved 13 May 2018

2

Methods for Food Safety

There are a number of technologies and methods which are used to ensure the safety of food. It includes hurdle technology, food irradiation, canning, freezing, pasteurization, biopreservation, etc. This chapter closely examines these different food safety methods to provide an extensive understanding of the subject.

Safe Food Handling Procedures

Handling food properly and safely is essential to preventing food borne illness.

- Preparing food in a safe manner.
- Serving food in a safe manner.
- Stopping the spread of bacteria through cross contamination.
- Routines to follow and habits to avoid.
- Presenting food in a hygienic and appetising way.

The importance of following proper safe food handling procedures.

From the time the food is delivered to the minute it is served to the customer, food safety should be at the top of the list.

Food business operators in particular should bear in mind that they are required by law, to ensure that any of their staff who handle food receive appropriate training in hygiene matters that are in line with their work activity.

There is little margin for error in any stage of food handling, whether it is preparation, processing, packaging, storage, transportation or offering it for sale. Also, note that if you prepare high risk foods the standards required of you will be much stricter than if you only prepare low risk foods.

- Protects people from getting sick.
- Protects your businesses reputation with customers.
- Protects your job.

The handling of food can take place during:

- Cooking,
- Cooling,
- Hot holding,
- Preparation,
- Purchase,
- Receipt,
- Re-heating,
- Serving,
- Storage.

General Safe Food Handling Tips

- Do not wipe your hands on your clothing as this can easily transfer microbes and bacteria.
- Use paper towels to clean up during food preparation and serving.
- Change gloves, utensils and dishes when changing functions. For instance use one pair of gloves for handling raw meat, and another pair handling fresh vegetables.
- Never run in food production or service areas.
- Try to have just one person serve food that is about to be eaten.
- Prepare precooked frozen foods exactly as the directions/instructions on the packaging state.
- Have foods ready not any longer than necessary before serving time.
- Prepare and cook only as much food as you intend to use.
- Wash and sanitize flatware or other utensils, which fall to the floor.
- Do not taste foods with any utensil used either to mix or stir food.
- Pick up and hold all tableware by the handles.
- Store tableware away from dust.
- Be careful when lifting lids from hot food.
- Turn handles of saucepans away from the front of the stove when cooking.

Picking up Ready to Eat Food

Whenever possible always try to handle any food items that are about to be eaten, with a utensil (i.e. tongs) rather than your bare hands.

Hand Washing

Clean hands are essential for working in a kitchen environment. It's very easy for bacteria to spread from the food we touch to door handles, plates, cutlery and so on. Hand washing is one of the best ways to prevent the spread of germs between people.

Use a soap dispenser rather than a bar of soap:

- Wash in a sink that has hot and cold running water.
- Wash in a sink that is separate from one that is used to wash foodstuff and utensils.
- Dry your hands with paper towels.

Wash your hands after:

- Starting work.
- Using the toilet.
- Handling raw and cooked foods.
- Taking breaks.
- Eating.
- Drinking.
- Smoking.
- Coughing, sneezing or blowing their nose.
- Touching your hair.
- Playing with pets or handling animals.
- Scratching.
- Handling refuse or waste materials.
- Handling cleaning chemicals.

Procedure to washing your hands properly:

- Wet your hands.
- Rub your hands and wrists with soap.
- Lather the soap for 20 seconds.

- Rinse thoroughly.
- Dry with paper towels or a hot air dryer (remember that wet hands can carry and transfer more germs than dry ones).
- Turn of the taps with your elbows (if possible) or use a paper towel to do so.

Hand Basins and Sinks

The sink you wash your hands in should be separate from ones where you prepare food or washing dishes. It should be in an accessible place, as this encourage people to use it and make it more likely to be used.

Gloves

Gloves are ideal for helping you to minimize bare hand contact with any cooked and ready-to-eat foods. They are there to protect both the food and the worker (i.e. they can be used to cover damaged skin or protect hands from risk of developing skin conditions).

Gloves must not be regarded as a "second skin". They can become contaminated with bacteria in exactly the same way that hands can. They are not a substitute for good personal hygiene and hand washing.

- Replace gloves after each task.
- Wash and dry hands thoroughly before putting on any gloves.
- Always use single use fresh gloves.
- Throw away plastic gloves after one use.
- The improper use of gloves can increase rather than reduce food hygiene risks, for instance a punctured glove can lead to glove material ending up in food.
- Gloves must only be used for one particular task.

Change gloves:

- At least once every hour.
- If they become contaminated.
- If they tear.
- When switching between handling raw and ready-to-eat foods.
- When changing tasks.
- After mopping, taking rubbish out, sweeping and cleaning.

Handling Dishes, Crockery and Cutlery

- Try not to touch any part of a dish or plate which will come into contact with a person's food or mouth.
- Pick up cups and mugs by their handles, your fingers should be outside cups.
- Place teaspoons so they protrude from a dish.
- Pull out disposable cups from the base of a tube, this prevents your fingers from going inside the cup.
- Do not use plates which have become cracked or chipped.

Clothes

Try to avoid wearing outdoor clothes in a food preparation area, instead wear clean, and where appropriate, washable protective clothing.

Wear:

- A clean apron.
- Gloves.
- Hairnet.
- Closed-in shoes to protect your feet, in case of hot spills or breakages.
- Shoes with slip-resistant soles, to stop you from slipping on hot spillages, etc.

Do not:

- Use your apron to wipe your hands on.
- Cook in loose fitting clothes.
- Work in the kitchen in soiled clothing.

Personal Hygiene

Food service workers must maintain a high degree of personal cleanliness when receiving, storing, cooking, processing, packaging, transporting or disposing of food. Here are some basic tips to follow:

- Keep fingers away from your face, mouth, hair, skin and other parts of the body.
- Don't brush or comb your hair when you are near food.
- Wash your hands frequently.
- Never smoke in food areas.

- Do not handle food with bare hands – use gloves instead.
- Do not eat or chew gum in food handling areas.
- Don't cough, sneeze, spit or smoke near food and avoid touching your nose, teeth, ears and hair, or scratching when handling food.
- Do not use fingers to sample food. Always use a clean spoon.

Using Knives

Always handle knives and other sharp equipment with care. Accidents involving knives are common in the catering industry, and usually involve cuts to a person non-knife hand and fingers. When using a knife always:

- Cut away from yourself or downwards on a chopping board to avoid cutting yourself.
- Cut on a stable surface.
- Keep knives clean, sanitised and grease free, all of these will help you have a firmer grip.

Tips:

- Use a knife suitable for the task and for the food you are cutting.
- Keep knives sharp.
- Carry a knife with the blade pointing downwards.

Using a Knife

When using a knife remember to focus on your:

- Stance or posture.
- Grip on the handle.
- Guiding or free hand.

Do not:

- Leave sharp knives loose in a drawer.
- Put knives in the sink.
- Use a knife as a can opener.
- Carry knives while carrying other objects.
- Engage in horseplay with a knife.

- Carry a knife in your pocket.
- Run your fingers down the edge of a knife to test the sharpness.
- Attempt to catch a falling knife.
- Put in the dishwasher.

Washing Knives

To prevent rusting and cross contamination, always wash and dry your knife immediately after you have finished using it. Do not let knives soak, especially if they have wood handles as the wood can expand when soaked in water.

Storing your Knives

Store them in a special knife rack or wooden block. This way you can help keep the blades sharp by keeping the edges away from hard objects that can dull the blades.

Hot Holding and Cold Holding Food

If you are holding foods for service, such as on a buffet line or in a cafeteria, then try to keep hot foods hot and cold foods cold. Hot holding equipment along with chafing dishes, slow cookers, and warming trays all help to keep ready to eat food out of the danger zone. All of this equipment is for hot holding only, and should not be used to reheat or cook food.

Tips:

- Preheat hot holding equipment before you put any food in it. If you don't then you'll be putting food into cold equipment which encourage bacteria growth.
- Limit the hot holding of food to a maximum of two hours.
- To distribute the heat evenly, make sure to stir the food at regular intervals.
- Keep the food covered, this not only retains the heat but also stops contaminates from falling into the food.
- Bring out the food as close as possible to the time of service.
- Keep platters refrigerated until it is time to warm them up for serving.

Pot Handles

Turn pot handles away from the front of the stove. This stops children from grabbing them, and adults from accidentally bumping into them.

Perishable Foods

After, a delivery always unload perishable foods first and immediately refrigerate them.

Using Kitchen Appliances

- Make sure that all necessary guards are in place before operating any equipment.
- Do not distract a colleague who is operating dangerous kitchen appliances like mincers or mixers etc.
- Do not to operate any machinery or use any chemical until it has been assessed by a qualified person.
- Make sure you are properly trained to use any kitchen appliances.
- Wash and put away appliances that are not being used, do not leave them lying around.
- Return equipment to it's correct storage place or location.
- Turn off all equipment and appliances at the end of each shift.

Children and Non Food Workers

Do not allow children, and people not involved in any cooking to roam or loiter around a food preparation area.

Work Surfaces

Make sure that work surfaces and equipment are visually clean, this goes a long way towards ensuring that they are free from high levels of harmful bacteria.

Clean as you Go

Train yourself to 'clean as you go', for instance cleaning up any spillages immediately.

Cans

Before opening a can of food always clean the top of it first. Remember that once the can is opened, any food which is not used immediately must be quickly stored in food grade containers and placed in a refrigerator.

Can Openers

Food can be left on any can opener after it has been used, it's therefore advisable to clean it after each use.

Plates

Never place cooked food on a unwashed plate that had previously held raw meat, poultry, or seafood.

Food Labels

Take the time to read product labels very carefully, and look for advisory statements like 'may contain ingredient X'.

Ovens

Close oven doors straight after removing or adding food items.

Meat and Poultry

Keep meat and poultry in its packaging until just before using.

Towels and Sponges

- Replace and wash dish towels and sponges often to prevent the spread of harmful bacteria throughout the kitchen.
- Do not use damp cloths when lifting hot items of equipment.

Uncovered Food

Try not to leave food unattended or uncovered for long periods.

Cutting Boards

Use separate cutting boards, dishes, utensils and cooking equipment for vegetables, raw meat and cooked meats.

Plates

When handling plates and trays do not touch eating surfaces with fingers.

Unused Sauces

Keep unused condiments, marinades and sauces separate from leftover ones.

Storing Food in the Fridge

Store raw meat, poultry and seafood by tightly wrapping it and then placing it on the bottom shelf of a refrigerator. This basically prevents the raw juices from dripping on other food.

- Refrigerate or freeze perishables, prepared food and leftovers within 2 hours.

Jewellery

Do not wear any watches, rings, bracelets or other jewellery when working with food. Germs can hide under them or just as worse they could accidentally fall off into the food.

Mitts

Use oven mitts when taking hot dishes from an oven or microwave. Do not use a wet oven mitt, as it can present a scald danger if the moisture in the mitt is heated.

Steps for Maintaining Food Safety

Keep the refrigerator temperature at 4 degrees Celsius. This temperature is intended to prevent the development of most pathogenic bacteria that may be in food.

Cool food down within a short time after cooking or consuming it to 4 degrees Celsius. Avoid keeping food outside of the refrigerator for more than an hour. Cooked food should be cooled down by various means, such as by immersing in a container with cold water, within two hours of cooking it, to the temperature of the refrigerator.

It is recommended to use disposable paper towels in the kitchen. If non-disposable tea towels are used, they should be washed frequently. Towels that are used to clean working surfaces that have been in contact with food could become contaminated with bacteria, some of which are pathogenic bacteria that convey diseases.

Cutting boards that come into contact with raw meat, fish and poultry should be cleaned with soap and hot water after each use. The cutting board should be rubbed with a scrubber sponge or brush with dishwashing liquid, rinsed well with hot water and air-dried. The scrubber sponge should be replaced frequently. These measures are in order to prevent the development of bacteria that remain on the cutting board. Do not use the same board to cut cooked meat.

Raw food, especially red meat, poultry and fish products, should be cooked thoroughly. Cooking of food, including meat, must reach a minimum temperature at the center of the product of 72 degrees Celsius. Meat products that have undergone cooking change their color. It is recommended to use a special kitchen thermometer, a special thermometer probe for checking food to measure the temperature at the center of the food and ensure thorough cooking. Do not use a mercury thermometer made of glass.

Do not consume raw eggs or eggs that have been treated with mild heat (soft boiled egg, fried egg). Do not prepare foods at home that to not undergo heat treatment and are based on fresh eggs, e.g. ice cream, mayonnaise, etc. Foods prepared with fresh eggs that are not destined to undergo heat treatment could be contaminated with Salmonella enterica bacteria (a bacteria that causes food poisoning).

Clean kitchen utensils (cutlery, plates, pots) with a scrubbing sponge or brush, hot water and dishwashing fluid; and air-dry them after rinsing in water. Avoid wiping kitchen utensils with a tea towel which could be contaminated with bacteria and re-contaminate the clean food utensils.

Wash your hands with hot water and soap immediately after handling raw food: vegetables, fruit, raw meat, poultry, fresh eggs or fish.

Defrosting of frozen food should be done in the refrigerator or microwave oven in a suitable container that prevents direct contact with other food products. Defrosting of uncooked meat, fish and poultry shall be done in the refrigerator only, on the lowest shelf.

Do not purchase perishable foods (meat, poultry, fish, dairy and egg products) at places that are not organized, air-conditioned, and have appropriate equipment such as refrigerators and freezers.

When purchasing fruits and vegetables, avoid their direct contact with other food products. Do not eat fruits or vegetables before cleaning them.

Cleaning fruits and vegetables:

- Wash the fruit/vegetable well with tap water, until the gross debris is removed.
- Soak the fruit/vegetable in water with liquid soap, and rinse using a brush.
- Dry the fruit/vegetable in air or with a clean paper towel.

ISO 22000

ISO 22000 is an international standard that defines the requirements of a food safety management system covering all sizes of all organizations throughout the food chain.

ISO is a global association that's made up of various national standards bodies. Together with its members and both government and non-governmental international organisations, ISO has developed the international standards for food safety. Since food safety incidents can happen at any stage of the food industry supply chain - manufacturing, transporting or selling - it's essential to have an effective food safety plan in place.

The ISO 22000 standards outline exactly what needs to be in a Food Safety Management System (FSMS). This is to make sure that food is kept safe during the entire food supply chain, right up until when the food is eaten. The standards are made up of these key parts:

- System management.
- Interactive communication.
- Prerequisite programs.
- HACCP principles.

ISO has listed the following requirements for a Food Safety Management System:

- To plan, implement, operate, maintain and update a Food Safety Management System that's aimed at providing products that are safe for the consumer.
- To comply with the necessary statutory and regulatory food safety requirements.
- To evaluate and assess the needs of customers and show that it conforms with the mutually agreed upon customer needs relating to food safety, which aims to improve customer satisfaction.
- To effectively communicate food safety issues to their suppliers, customers and relevant interested parties throughout the food supply chain.
- To make sure that the organisation complies with its stated food safety policy and effectively shows this.
- To seek certification or registration of its Food Safety Management System by an external organisation, or make a self-assessment or self-declaration of conformity to ISO 22000:2005.

Need for ISO 22000

Every day all around the world food is produced, transported, sold and consumed. This means there's always a risk of contamination and other food safety hazards. The ISO

22000 standards aim to lower the risk of bacterial contamination and therefore the risk of disease throughout the entire food supply-chain process.

ISO 22000 provides businesses with a comprehensive framework for a Food Safety Management System related to food safety. An organisation can purchase different ISO 22000 packages depending on their individual requirements.

Organisations certified in ISO 22000 can indicate to their customers that they have a Food Safety Management System in place. Having an ISO 22000 certification can help make customers feel more secure in the safety of a business's product; an increasingly important issue for both businesses and consumers.

International food safety standards are continually reviewed so as to remain relevant to the current food safety industry. The next update of the ISO food safety standards is expected to be published in 2017.

Working of ISO 22000

ISO 22000 covers the entire food industry supply-chain – from farm to plate and outlines the relevant procedures required to meet the various food safety standards.

ISO 22000 integrates the HACCP principles; interactive communication; system management; and prerequisite programs.

HACCP Principles

ISO 22000 reflects the principles of the Hazard Analysis and Critical Control Point (HACCP) system. Analysing potential hazards is a key element of maintaining a successful Food Safety Management System as it provides the organisation with effective measures of control.

ISO 22000 requires an organisation to assume that any reasonable hazard is expected to occur, whether it be associated with the organisation's equipment or procedures, and is consequently identified and evaluated.

Interactive Communication

The ISO 22000 standards outline that communication is imperative throughout all stages of the food supply chain. Effective communication ensures that all potential food safety hazards are properly recognised and therefore, suitably controlled.

Communication is essential not just within an organisation, but also with both customers and suppliers.

System Management

A properly structured management system is essential to ensure that a Food Safety

Management System is correctly carried out in an organisation. It must be properly outlined, maintained and updated within the organisation's structured management system. This makes certain that the Food Safety Management System is of maximum benefit to the organisation itself and all other interested parties.

Prerequisite Programs

Prerequisite programs are policies and procedures implemented in an organisation that relate to the role the manufacturing environment plays in creating safe food products. ISO 22000 requires an organisation to identify and implement the necessary prerequisite programs in order to control the risk of contamination.

ISO 22000 does not require particular prerequisite programs, unlike some other certified Food Safety Management Systems. Rather, it requires an organisation to recognise and apply the required programs specific to its individual needs.

ISO 22000 can be integrated into an existing related management system or applied independently of other management system standards. Organisations can also establish a new Food Safety Management System and ensure that it meets the requirements of the ISO 22000 standards.

Benefits of ISO 22000

- Introduce internationally recognized processes to your business.
- Give suppliers and stakeholders confidence in your hazard controls.
- Put these hazard controls in place across your supply chain.
- Introduce transparency around accountability and responsibilities.
- Continually improve and update your systems so it stays effective.
- ISO 22000 contains the food safety management system requirements of FSSC 22000 (which is a Global Food Safety Initiative, GFSI recognised scheme) and is used along with requirements for prerequisite programs for the appropriate industry sector.

Hurdle Technology

Most of the food products in the market are preserved (i.e. retained its stability) based on more than one hurdle or preservation method. In order to determine the food stability, two questions need to be asked: what target attribute(s) needs to be achieved in the microbial, chemical, bio-chemical and physical changes; and what is the required time frame of stability? When considering stability, the microbial and chemical safety aspects must be considered first before sensory properties. The microbial stability and safety of the most traditional and novel foods is based on a combination of several preservation factors (called hurdles), and the microorganisms present in food are unable to overcome. This is illustrated by the so-called hurdle effect, first highlighted by. The critical limits are being used by the industry when each hurdle such as heat treatment, water content, pH and storage temperature is applied alone. Fundamental based theoretical concepts of F-value (hurdle: Heat treatment), water activity (hurdle: Water content) and glass transition (hurdle: Glassy state; depending on water, storage temperature, and structure) are the most successful in determining food stability during food processing and storage. These concepts (i.e. each hurdle) are usually applied to specific types of products, for examples F-value to canned foods (i.e. high moisture) and water activity and glass transition for dried and frozen foods. The F-value is based on commercial sterility, water activity by state of water (i.e. bound or free) and glass-rubber transition by structural mobility. However, more than 60 hurdles may involve in food preservation.

In achieving the desired safety by only one hurdle, high severity in processing needs to be applied. This caused significant damages to the nutritional and sensory quality of foods. For this reason, it is important to have multi-hurdles approach for developing safe and wholesome food products. The hurdle effect has fundamental importance for the preservation of foods, since the hurdles provide a control to microbial spoilage, food-poisoning and other undesired changes. The advantages of hurdle concepts are (1) it can avoid the severity of one hurdle for preservation, (2) it can give synergy of combination, and (3) many of the hurdles come from past experience (i.e. tradition or culture). Currently huge numbers of products are being developed based on the multi-hurdles. The combinations and the levels of hurdles are determined based on the empirical experiments. However, it is a challenge to food scientists and engineers to have unified concept or approach for determining food stability considering multi-hurdles, such as heat treatment, water content, pH, salt, spices, preservatives, packaging and storage temperature. Stability map was proposed based on the state diagram (i.e. states and phases of a food as a function of water or solids content and temperature). Recently macro-micro region concept in the state diagram has been proposed and relative stability map is postulated in the 13 micro-regions. In order to achieve safety, the proposed micro-region concept showed potential to combine multi- hurdles, or to provide a guide on the hurdles need to be used in each micro-region. The objective of this chapter is to highlight the recent development in achieving food stability by intelligent use of multi-hurdles and theoretical concepts.

Guide Lines for pH and Salt

pH Value

The minimum pH for the growth and toxin production of microorganisms are compiled by. Here we include a pH lower than 4.6 to inhibit the growth of many pathogens and this limit is considered as safe or low risk. It is generally accepted that the limiting pH of 4.6 provides a good margin of safety against the hazards of botulism in acidified foods, and such products are given relatively mild heat treatment. The critical limit of pH 4.6 is commonly used to categorize the low risk and high risk foods. the IFT panel concluded that a pH of 4.6 is appropriate to control spore-forming pathogens and a pH of 4.2 is appropriate to control vegetative pathogens.

Salt (i.e. Sodium Chloride)

Salt is one of the most common preservatives. Life at high salt concentration is energetically expensive. identified thermodynamic limits to microbial life at high salt concentration. They identified that 346 g/L (i.e. 34.6 %) was the limit to stop all microbial process, thus only salt as a hurdle could be impossible to implement due to the negative impact on health and sensory. For this reason, most of the cases salt is used in combination with other hurdles. Most processed meat products contain sodium chloride from 2.8 % in cooked sausages to 4.5 % in cured meat products and have been maligned as a heavy salt contributor to the diet. However, in many products salt in combination with other hurdles showed high potential.

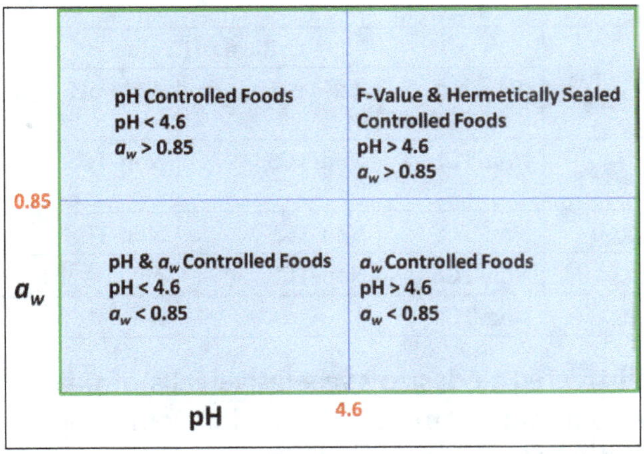

FDA Good Manufacturing Practice Regulations governing processing requirements and classification of foods.

The stability guidelines are mainly based on the pH, water activity, and thermal processing. FDA's Good Manufacturing Practice Regulations governing the processing requirements and the classification of foods are shown in figure. Low acid (i.e. high pH) foods packaged in hermetically sealed containers must achieve commercial sterility conditions either by retorting or combined treatment of pasteurization and water activity or a combined treatment of pasteurization and acidification. It could be seen that pH 4.6 and a_w 0.85 are the critical limits. In 1962 the US Public Health Service in the "Food Service Sanitation Manual" issued the potentially hazardous food (PHF) as any perishable food which consists in whole or in part of milk products, eggs, meat, poultry, fish or other ingredients capable of supporting the rapid and progressive growth of infectious or toxigenic microorganisms. The progress of the definition of PHF is discussed in the FDA

Food Code in 1999 defined PHF food as it must be kept cold or hot, because the food (i.e. temperature controlled for safety, TCS) has the necessary intrinsic factors to support the growth of pathogens. The TCS foods require time/temperature control to limit pathogen growth or toxin formation that constitutes a threat to public health. A science-based framework was proposed by the IFT/FDA panel. The framework contains two steps:

Table: Control of spores: product treated to control vegetative cells and protected from recontamination.

Critical a_w Values	Critical pH Values		
	pH ≤ 4.6	4.6 < pH < 5.6	p H 5.6
$a_w \leq 0.92$	Non-TCS	Non-TCS	?
$0.92 < a_w < 0.95$	Non-TCS	Non-TCS	?
$a_w > 0.95$	Non-TCS	?	?

Table: Control of vegetative cells and spores: product not treated or treated but not protected from recontamination.

Critical a_w Values	Critical pH Values			
	pH < 4.2	4.2 < pH < 4.6	4.6 < pH < 5.0	pH > 5.0
$a_w < 0.88$	Non-TCS	Non-TCS	Non-TCS	Non-TCS
$0.88 < a_w < 0.90$	Non-TCS	Non-TCS	Non-TCS	?
$0.90 < a_w < 0.92$	Non-TCS	Non-TCS	?	?
$a_w > 0.92$	Non-TCS	?	?	?

Step 1: Was the food treated to destroy vegetative cells of potentially pathogens and packaged to avoid recontamination? If yes, position your product in Table according to its pH and water activity (a_w). If not, position your product in Table according to its pH and aw.

Step 2: If the food is classified as a non-TCS food according to Step 1 above, it may be stored and held safely without regard to time or temperature. If the need for time/temperature control is questionable, the food should be held either hot or cold for safety or subjected to a product assessment as the next step in determining the appropriate classification.

The product assessment was performed based on the factors, such as ingredients, processing, change of acids used to lower pH, preservatives, and formulation. The final decision on the hurdles could be based on challenge testing and available predictive models. The panel used their proposed framework to determine its applicability to a specific example(s) from selected class of food product categories. Additional examples of the determination of TCS are provided in the report developed by Texas Department of State Health Services. They also identified the pathogens of concern for different classes of food products and potential hurdles alone or combination to be used. Scientific sound criteria for determining whether foods require time/temperature control for safety could consider the interaction of pH, aw and other factors from microbial growth models. For example, USDA pathogen model could identify the critical boundary for growth of specific bacteria when both pH and a_w are used. The panel pointed that framework should be validated for a broad variety of products, and predictive models of pathogens as a function of several parameters, such as packaging atmosphere, redox potential, pH, a_w, preservatives and ingredients should be developed. In addition, synergistic factors and variations of results in the real product and challenged test should also be explored.

Concepts of Food Stability Determination

F-value

In 1795 Nicolas Appart, a chef took up a challenge and established a theory that if fresh foods were put in airtight containers and sufficient heat applied, then food would last longer. He proposed his preservation method after 14 years of experimentation without understanding the role of bacterial spoilage. A theoretical understanding of the benefits of canning did not come until Louis Pasteur observed the relationships between micro-organisms and food spoilage after 50 years later. The sterilization of canned foods has a long tradition and it is most likely that it will continue to be popular because of its convenience and extended shelf-life (1–4 years at ambient temperature).

The time of sterilization process is estimated based on F-value concept. The F-value was originally developed by Bigelow and subsequently by Ball and Olson. By the 1860s the time required to process food in can was reduced from 6 h to 30 min through numbers of inventions and improvements Simpson. The inactivation target in sterilization process depends on the types of heat resistance bacteria and pH of foods. It is possible to classify food products into three groups according to pH: low-acid (pH ≥ 4.6), medium-acid (3.7 ≤ pH ≤ 4.6), and high-acid foods (pH ≤ 3.7). The target microorganism in the thermal processing of low-acid food (pH ≥ 4.6) is Clostridium botulinum. The processing time based on first order kinetics can be estimated as:

$$t = -\frac{1}{k}\ln\left(\frac{N_F}{N_O}\right) = \left(\frac{D}{2.303}\right)\ln\left(\frac{N_F}{N_O}\right),$$

Where k is the destruction rate (s−1), N_o and N_F are the initial and final numbers of micro-organisms, and D is the decimal reduction time, respectively. Equation for microbial death indicates that final concentration tends to zero when time tends to infinity, thus it would not be possible practically to reach a final concentration of zero for the target micro-organism. Considering this point, the commercial sterilization criterion should be defined so that it is possible to design a process that is safe but occurs within a finite time and which is economically and practically feasible. According to Stumbo the commercial sterilization criterion was established arbitrarily. The commercial sterilization criterion states that the minimum thermal process should reduce initial micro-organism concentration by 1012. This is well known as 12D concept or "botulinum cook". discussed the origins of the 12D concept and outlined the original work of. They concluded that was the first to recognize the logarithmic nature of time temperature curve. The commercial sterilization was used for several reasons (1) safety margins, (2) cooking requirements, and (3) to prevent the growth of thermophilic spoilage microorganisms. The probability argument says, in 12D treatment there will be one spore in 1012 cans. Considering 100 million cans consumed per day, an estimate over a 100-year period worldwide consumption will be 3.65 × 1012 cans and the 12D criterion would predict three to four outbreaks every 100 years.

The processing time to achieve 12*D* can be estimated when D_T value of target micro-organism at the temperature *T* is known. However, the processing time varies with the size of a can. Considering $D_{121.1}$ = 0.21 min for *C. botulinum*, one spore per g and can size of 0.1 L (density: 1 g/cm³), the processing time (t) to achieve 12*D* is 2.94 min from equation. Similarly if the can size is 5 L, the processing time is 3.29 min. Since thermal process could not be performed at a specific temperature instead a varied temperature range during heating and cooling time, thus thermal death time, F_r is defined for processing as:

$$F_r = \int_0^t 10^{\frac{T-Tr}{z}} dt,$$

Where F_r is the processing time at a reference temperature (s), T_r is the reference temperature (°C) and *z* is the thermal constant or *z*-value. If reference temperature T_r is 121.1 °C, then F_r is termed as F_o as:

$$F_o = \int_0^t 10^{\frac{T-121.1}{z}} dt,$$

Currently, a common commercial sterilization treatment for C. *botulinum* F_o is in the range of 6–8 min, although some companies use F_o of 10 min or higher. Considering this processing time, 5 L can size and the minimum time requirement (6 min) at 121.1 °C, indicated 1.335×10^{-25} spores per package. Applying the probability concept, we should expect one outbreak in several billion years (to be precise, one hundred thousand billion years). In fact, at least in the past 50 years no outbreak has been directly related to the sterilization criterion. However, there is no theoretical method available for the prediction of F_o, if other hurdles are used in combination.

Water Activity

W.J. Scott, an Australian scientist, proposed that the active water could be much more important to the stability of a food than the total amount of water present. The legacy of Scott allowed scientist to develop generalized rules or limits for the stability of foods using water activity. In general the rules of water activity concept are (1) food products are most stable at their "BET-monolayer" content or "BET-monolayer water activity" and unstable above or below BETmonolayer; (2) there are a critical water activity limit for a specific micro-organism or a class of micro-organism for their growth or toxin production, and biochemical reactions. For example, there is a critical water activity level below which no microorganisms can grow. Pathogenic bacteria cannot grow below a water activity of 0.85, whereas yeasts and molds are more tolerant to reduced water activity, but usually no growth occurs below a water activity of about 0.6. proposed the food stability map based on the water activity concept containing growth of micro-organisms and different types of bio-chemical reactions. In the recent food stability map, showed the trends of microbial growth, bio-chemical reactions and mechanical characteristics in the three zones of water activity (zone 1: BET-monolayer, zone 2: adsorbed multi-layer, zone 3: matrix or solvent water). In fact, the BET-monolayer could be only achieved in the cases of dried foods. However, food industries are now widely used this concept for determining the stability of their dried products.

Glass Transition

Considering the limitations of water activity concept, the hypothesis of glass transition concept was put forward. White and first highlighted the importance of the glassy state of foods in determining its structural stability. The significant applications of the glass transition concept emerged in food processing in the 1980s, when Levine and and Slade and identified its major merits and wide applications. The rules of the glass transition concept are (1) the food is most stable at and below its glass transition (i.e. T_g or T_g'), and (2) the higher the $T-T_g$ or T/T_g (i.e. above glass transition), the higher the deterioration or reaction rates. The limitations of water activity and glass transition concepts would not invalidate the concepts completely rather make it difficult to apply universally. The water activity concept is based on the binding nature of water molecules in the matrix.

When water is bound (i.e. unavailable to take part in reactions) to the solid matrix or non-solvent, then no deterioration reactions could be expected. The glass transition concept is based on the molecular mobility of the reacting components at micro-level in a matrix, thus diffusion of the reactants through the system is very slow and stability is achieved. Thus a successful combination of water activity and glass transition could open more precise and unified determination of stability criteria.

Critical Temperature Concept and Molecular Mobility

It is expected that there should be a break in the plot of k (i.e. reaction rate) versus T/T_g (i.e. change in slope between above and below the critical ratio) at T/T_g equal to 1, if glass transition concept is valid or X_w/X_b equal to 1, if water activity concept is valid. measured molecular mobility by ST-EPR and 1 H- NMR, and observed two distinct changes: first one minor shift just close to T_g and second abrupt decrease due to solid-like to liquid-like defined as T_c. In the case of sugars, T_c was observed at 17–35 °C higher than T_g and for biological materials it was more than 50 °C. This variation was also explained by the density of hydrogen bond and molecular packing measured by FTIR. This higher T_c was also correlated with the observed collapse or softening of sugars at 10–17 °C above glass transition and crystallization above 30 °C. It is generally believed that crystallization over practical time scale occurred above glass transition, although some report showed it was below 30 °C than glass temperature. α-amylase was more stable in rubbery matrices of lactose or trehalose than in a glassy PVP matrix and the protective efficiency of saccharides, maltodextrins and PVPs did not increase with their respective glass transition temperature. In addition dielectric and other spectroscopy determine α, β and γ relaxations below glass transition.

However it is not clear how these could be related or linked to determine the stability of foods. Considering the fact that glass transition is not the critical limit, Rahman tested the hypothesis that there is a critical temperature as a ratio of T_c/T_g (T_c is the critical temperature) which could vary with moisture content. Above the critical temperature, an increase in the water content or temperature significantly increased the reaction rate while below the critical temperature the rate was relatively less affected by water content and temperature. He observed values of T_c/T_g varied from 0.78 to 1.5 depending on the types of reaction and the matrices. In some instances, the values of T_c/T_g were close to 1.0 indicating only glass transition could explain the process. Moreover, the deviations of T_c/T_g from 1 explain why in many instances both stability and un-stability were observed above and below glass transition. The glass transition by thermal or mechanical relaxations measure mobility in a 20–300 nm range, while other relaxation techniques, such as Nuclear magnetic Nuclear Magnetic Resonance (NMR) measures the molecular relaxation in a 1–2 nm range Recently, tremendous progress has been achieved in understanding multidimensional aspects of the molecular mobility. Although it is not yet very clear how these knowledge could by applied university in determining the stability of foods.

State Diagram

The glass transition concept was further advanced by developing state diagrams for foods. A state diagram is a stability map of different states and phases of a food as a function of water or solids content and temperature. Most probably, Levine and Slade presented the first state diagram in the food science literature by illustrating glass line, freezing curve, and intersection of these lines as T_g'' by extrapolation of the extended freezing curve by maintaining similar curvature. The main advantage of drawing a map is to help understanding the complex changes when the water content and temperature of foods are changed. It also assists in identifying stability of foods during storage as well as selecting a suitable condition of temperature and moisture content for processing. The state diagram based on freezing curve and glass transition provided four macro-regions: Region I (i.e. below glass transition), region II (above glass transition and below maximal-freeze-concentration, i.e. completely frozen), region III (above maximal-freeze-concentration condition and below freezing curve, i.e. partially frozen), and region IV (i.e. above glass transition and above freezing curve). Rahman combined water activity and glass transition concepts in the state diagram by plotting BET- monolayer as a function of temperature. This makes four regions: below BET-monolayer, one above and one below; and above BET-monolayer, one above and another below. It was emphasized that a combination of water activity and glass transition concepts could be a powerful tool in predicting food stability. The approaches to combine both concepts are reviewed by Rahman. However, Rahman combined glass transition and water activity concepts in the state diagram by plotting BET-monolayer values as a function of temperature.

Macro-micro Region in the State Diagram

Using state diagram, Rahman hypothesized 13 micro-regions having the highest to the lowest stability based on the location from the glass and BET-monolayer lines. For example, region-1 (relatively non-reacting zone, below the BET-monolayer line and glass line) is the most stable and region-13 (highly reacting zone, far from BET-monolayer line and glass line) is the least stable. The stability decreased as the zone number increased. Applications of this hypothesis in food processing are presented by Rahman. Schebor. studied sucrose hydrolysis in the micro-regions 1 and 2, and evidenced the validity of the hypothesis. Advantages of micro-region concept are as follows (1) stability rules could be developed for each micro-region (i.e. narrow moisture and temperature range) as compared to the macro-region (i.e. broad moisture and temperature region), and (2) the states or phases of the material could be identified in each micro-region. A reference point could be identified where BET-monolayer line and glass line intersect and any location in the state diagram could be assessed in relation to the reference point.

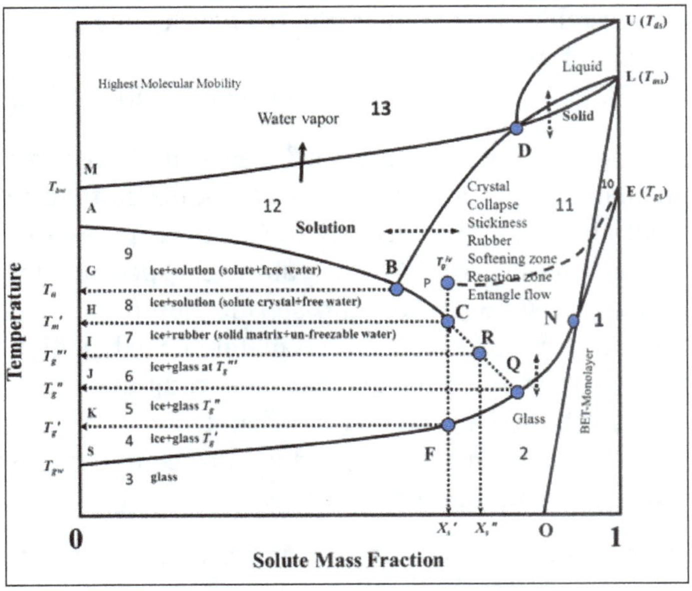

State diagram showing different regions and state of foods T_{ds}: Solids-decomposition temperature, T_{ms}: Solids melting temperature, T_{gs}: Solids–glass transition temperature, T_g: End of solids-plasticization temperature, T_{gw}: Glass transition of water, T_u (solute crystallization temperature during freeze-concentration), T_m' (maximal-freeze concentration condition, i.e. end point of freezing), T_g''' (glass transition of the solids matrix in the frozen sample as determined by Differential Scanning Calorimetry (DSC)), T_g'' (intersection of the freezing curve to the glass line by maintaining the similar curvature of the freezing curve), and T_g' (glass transition at maximal-freeze-concentration, i.e. at the end point of freezing), T_{bw}: Boiling temperature of water.

Progress in Microbial Reaction in Relation to Hurdle Concept

Hurdles are deliberately combined to improve the microbial and chemical stability, and the sensory quality of foods as well as their nutritional properties and economic cost. Thus it improves the total quality of foods by application of an intelligent mix of hurdles. The main advantage of the hurdle technology is its synergistic effect (i.e. non additive). Previously hurdle technology (i.e., a combination of preservation methods) was used empirically without much knowledge of the governing principles. Progress in principles of the major preservative factors on the physiology and behavior of microorganisms in foods could allow more intelligent applications of hurdles. These are homeostasis, metabolic exhaust, and stress reactions.

Homeostasis is the tendency to achieve uniformity and stability in the internal status of organisms. For instance, the maintenance of a defined pH is a prerequisite and feature of living cells, and this applies to higher organisms as well as microorganisms. When homeostasis of bacteria is altered, the bacterial cells react by spending their energy in maintaining their physiological status rather than in multiplying, thus repair needs more energy. In this case microorganisms remain in the lag phase or even die before their homeostasis is reestablished by repair. Energy restrictions for microorganisms are caused by anaerobic conditions, such as vacuum or modified atmosphere and low a_w, pH and redox potential, and these act synergistically in combination when applied in foods.

Metabolic exhaustion is the "auto-sterilization". It was first observed that in a mildly heated sausage inoculated with Clostrial spores and adjusted to different water activity by addition of salt and fat, and stored 37 °C. Clostrial spores surviving during the heat treatment vanished in the product during storage, especially unrefrigerated. Further experimental evidence showed that auto-sterilization occurred in the unrefrigerated conditions, and more hurdles accelerated metabolic exhaustion. More examples of "auto-sterilization" are presented by Leistner. The traditional "air-dried" fermented sausages from Germany showed a good record of safety due to the fermentation at low temperature (<15 °C) and to extensive aging, which was compensated the requirement of higher levels of acids, and antimicrobial nitrite.

Some bacteria become more resistant or even more virulent under stress (i.e. stress reaction), since they generate stress shock proteins. The synthesis of protective stress shock proteins is induced by heat, pH, a_w, ethanol, and oxidative compounds as well as by starvation. Therefore, multi-target preservation of foods could be the key to avoiding synthesis of stress shock proteins, which otherwise could jeopardize the microbial stability and safety of hurdle technology. Nevertheless, further research in stress shock proteins and different mechanisms that switches them on or could inactivate them warranted, in relation to hurdle technology.

The multi-target preservation of foods should be the ultimate goal for a gentle but most effective preservation of foods. Proposed that food microbiologist could learn from pharmacology and medicine. For example, at least 12 classes of biocides are already known which have different targets, and sometimes more than one, within microbial cell. Often the primary target is the cell membrane and it becomes leaky due to the disruption of cell wall.

In addition, biocides also impair synthesis of enzymes, proteins, and DNA. Multi-drug attack has proven successful in the medical field to fight bacterial infections (i.e. tuberculosis) as well as viral infections (i.e. aids), thus a multi-target attack on microorganisms should also be a promising approach in food microbiology.

Leistner provided an example of hurdle technology in the case of Indian paneer. It is a traditional cottage-cheese-type product fried in cubes with oil and onions and sauce containing salt, spices, and often tomatoes are added. However, paneer spoils bacteriologically within 1–2 days at room temperature (which in India can reach 35 °C), and this is a strong drawback for its industrial production. Sterilized paneer in cans showed severe sensory limitations with regard to flavour, texture and color. Table shows how hurdle technology can use to develop paneer with long shelf life. Another example on the La Chang (i.e. a meat product in China) is also presented by Leistner and the combinations of hurdles and the shelf life are shown in Table:

Table: Examples of hurdle products.

Product	Hurdles	Shelf-life
La Chang (meat) China	a_w: 0.85–0.70	2–3 months
	pH: 5.9–5.7	
	NaCl: 3–5 %	
	Sugar: 4–20 %	
	Refrigeration: No	
	Total count < 10^6/g	
La Chang (meat) China	a_w: 0.85–0.70	4–5 months
	pH: 5.9–5.7	
	NaCl: 3–5 %	
	Sugar: 4–20 %	
	Packaged: vacuum	
	Refrigeration: No	
	Total count < 10^6/g	

Fresh Paneer (India)	Storage: 35 °C	1–2 days
Paneer (India)	a_w: 0.97	Several weeks
	pH: 5.0	
	F_0: 0.8 min	
	Storage: 35 °C	
Paneer in gravy (India)	a_w: 0.95	2 weeks (Storage: 45 °C)
	pH: 5.0	1 month (Storage: 30 °C)
	F_0: 0.8 min	3 months (Storage: 15 °C)
	Potassium sorbate: 0.1 %	

Prediction for Multi-hurdles

In predicting or modelling stability, two approaches could be used in identifying the boundary or limit of the growth/no-growth (reaction or no-reaction) within a time frame, and then predicting the rate of growth or reaction within the growth or reacting region. developed a mathematical model to describe the logarithm nature of thermal death. Nevertheless it was not until the 1980s when mathematical models predicting the behavior of microorganisms experienced a great development. A new field emerged in the food microbiology area: Predictive microbiology. Different types of empirical growth models for microorganism are widely used. These are specific to the types of microorganism, physico-chemical properties of food, and the used hurdles. developed thermal inactivation rate (i.e. ln k) of Cl. Botulinum as a linear function of $1/T$, pH and pH^2. Similarly Cerf also developed the regression correlations of ln k as a function of 1/T, pH, pH^2, and a_w. The non-monotonous changes of pH and aw make it more difficultly to predict the effect of hurdles in combination. generated growth data of Salmonella in sterile pork at various isothermal conditions, and dynamic models were developed within storage temperature (10–45 °C).

In 1970s the "probability model" for the prediction of germination of spores, population growth, survival, and toxin production within a specified period of time under defined conditions of storage and product composition was emerged. The "growth/no-growth" or "interface" modeling are progressed. One of the first attempts to growth/no-growth (G/NG) modeling was developed by Pitt. He related the boundary by an empirical equation with the temperature and water activity limits for Aspergillus spp. growth. reviewed the methodology of development of a probability or growth/no-growth (G/NG) models. The logistic regression approach has been widely adopted for probability and G/NG modelling. In general, probability models are devoted to the data which can be measured as "positive" or "negative". For example, if we consider the variable "detection of toxin", only two responses are possible: "detectable" or "not detectable" and responses can be coded as 1 (positive response) or 0 (negative response), or even better, if response replicates are obtained, a number between 0 and 1 can be given to the response, calculated as the average of the scores (0 or 1) of the replicates. The number obtained is considered as the probability of occurrence of the phenomenon studied. This probability can be related to independent variables such as temperature or pH by some mathematical function using regression techniques. In logistic regression, the probability (*P*) can be expressed as "logit" as:

$$logit\,P = \log\left(\frac{P}{1-P}\right),$$

Mertens developed G/NG model by linear logistic regression and logit was correlated as a function of aw, pH, sodium chloride and acetic acid. A simplified G/NG model conceptually derived from the Gamma model. Logit was related to the variables or factors in order to determine the boundary. Developed G/NG boundary using gamma function and observed a limited numbers of fail dangerous prediction, thus he suggested simplified G/NG model could be used as a first estimate method when experimental data are missing or insufficient. In addition, a practical time frame usually used to generate the boundary data, thus it ignores stability for longer period. In general, extrapolation in prediction, especially in foods, is not straightforward because of the complexity of the food matrix.

Food Irradiation

Irradiation does not make foods radioactive, compromise nutritional quality, or noticeably change the taste, texture, or appearance of food. In fact, any changes made by irradiation are so minimal that it is not easy to tell if a food has been irradiated.

Food irradiation (the application of ionizing radiation to food) is a technology that improves the safety and extends the shelf life of foods by reducing or eliminating microorganisms and insects. Like pasteurizing milk and canning fruits and vegetables, irradiation can make food safer for the consumer. The Food and Drug Administration (FDA) is responsible for regulating the sources of radiation that are used to irradiate food. The FDA approves a source of radiation for use on foods only after it has determined that irradiating the food is safe.

Reason for Irradiating Food

Irradiation can Serve many Purposes

- Prevention of Foodborne Illness – To effectively eliminate organisms that cause foodborne illness, such as *Salmonella* and *Escherichia coli* (*E. coli*).
- Preservation – To destroy or inactivate organisms that cause spoilage and decomposition and extend the shelf life of foods.

- Control of Insects – To destroy insects in or on tropical fruits imported into the United States. Irradiation also decreases the need for other pest-control practices that may harm the fruit.
- Delay of Sprouting and Ripening – To inhibit sprouting (e.g., potatoes) and delay ripening of fruit to increase longevity.
- Sterilization – Irradiation can be used to sterilize foods, which can then be stored for years without refrigeration. Sterilized foods are useful in hospitals for patients with severely impaired immune systems, such as patients with AIDS or undergoing chemotherapy. Foods that are sterilized by irradiation are exposed to substantially higher levels of treatment than those approved for general use.

National Aeronautics and Space Administration (NASA) astronauts eat meat that has been sterilized by irradiation to avoid getting foodborne illnesses when they fly in space.

Sources of Food Irradiation

There are three sources of radiation approved for use on foods.

- Gamma rays are emitted from radioactive forms of the element cobalt (Cobalt 60) or of the element cesium (Cesium 137). Gamma radiation is used routinely to sterilize medical, dental, and household products and is also used for the radiation treatment of cancer.
- X-rays are produced by reflecting a high-energy stream of electrons off a target substance (usually one of the heavy metals) into food. X-rays are also widely used in medicine and industry to produce images of internal structures.
- Electron beam (or e-beam) is similar to X-rays and is a stream of high-energy electrons propelled from an electron accelerator into food.

Benefits of Irradiated Food

The FDA has evaluated the safety of irradiated food for more than 30 years and has found the process to be safe. The World Health Organization (WHO), the Centers for Disease Control and Prevention (CDC) and the U.S. Department of Agriculture (USDA) have also endorsed the safety of irradiated food.

The FDA has approved a variety of foods for irradiation in the United States including:

- Beef and Pork,
- Crustaceans (e.g., lobster, shrimp, and crab),
- Fresh Fruits and Vegetables,
- Lettuce and Spinach,
- Poultry,
- Seeds for Sprouting (e.g., for alfalfa sprouts),
- Shell Eggs,
- Shellfish- Molluscan (e.g., oysters, clams, mussels, and scallops),
- Spices and Seasonings.

Pasteurization

The process of pasteurization was named after Louis Pasteur who discovered that spoilage organisms could be inactivated in wine by applying heat at temperatures below its boiling point. The process was later applied to milk and remains the most important operation in the processing of milk.

The heating of every particle of milk or milk product to a specific temperature for a specified period of time without allowing recontamination of that milk or milk product during the heat treatment process.

Purpose There are two distinct purposes for the process of milk pasteurization:

- Public Health Aspect - To make milk and milk products safe for human consumption by destroying all bacteria that may be harmful to health (pathogens).
- Keeping Quality Aspect - To improve the keeping quality of milk and milk products. Pasteurization can destroy some undesirable enzymes and many spoilage bacteria. Shelf life can be 7, 10, 14 or up to 16 days.

The extent of microorganism inactivation depends on the combination of temperature and holding time. Minimum temperature and time requirements for milk pasteurization are based on thermal death time studies for the most heat resistant pathogen found in milk, *Coxelliae burnettii*. Thermal lethality determinations require the applications of microbiology to appropriate processing determinations.

To ensure destruction of all pathogenic microorganisms, time and temperature combinations of the pasteurization process are highly regulated.

Ontario Pasteurization

Milk:

- 63 °C for not less than 30 min.,
- 72 °C for not less than 16 sec.,
- Equivalent destruction of pathogens and the enzyme phosphatase as permitted by Ontario Provincial Government authorities. Milk is deemed pasteurized if it tests negative for alkaline phosphatase.

Frozen dairy dessert mix (ice cream or ice milk, egg nog):

- At least 69 °C for not less than 30 min;
- At least 80 °C for not less than 25 sec;
- Other time temperature combinations must be approved (e.g. 83 °C/16 sec).

Milk based products- with 10% mf or higher, or added sugar (cream, chocolate milk, etc) 66 °C/30 min, 75 °C/16 sec.

There has also been some progress with low temperature pasteurization methods using membrane processing technology.

Thermal Destruction of Microorganisms

Heat is lethal to microorganisms, but each species has its own particular heat tolerance. During a thermal destruction process, such as pasteurization, the rate of destruction is logarithmic, as is their rate of growth. Thus bacteria subjected to heat are killed at a rate that is proportional to the number of organisms present. The process is dependent both on the temperature of exposure and the time required at this temperature to accomplish to desired rate of destruction. Thermal calculations thus involve the need for knowledge of the concentration of microorganisms to be destroyed, the acceptable concentration of microorganisms that can remain behind (spoilage organisms, for example, but not pathogens), the thermal resistance of the target microorganisms (the most heat tolerant ones), and the temperature time relationship required for destruction of the target organisms.

The extent of the pasteurization treatment required is determined by the heat resistance of the most heat-resistant enzyme or microorganism in the food. For example, milk pasteurization historically was based on Mycobacterium tuberculosis and Coxiella burnetti, but with the recognition of each new pathogen, the required time temperature relationships are continuously being examined.

A thermal death curve for this process is shown below. It is a logarithmic process, meaning that in a given time interval and at a given temperature, the same percentage of the bacterial population will be destroyed regardless of the population present. For example, if the time required to destroy one log cycle or 90% is known, and the desired thermal reduction has been decided (for example, 12 log cycles), then the time required can be calculated. If the number of microorganisms in the food increases, the heating time required to process the product will also be increased to bring the population down to an acceptable level. The heat process for pasteurization is usually based on a 12 D concept, or a 12 log cycle reduction in the numbers of this organism.

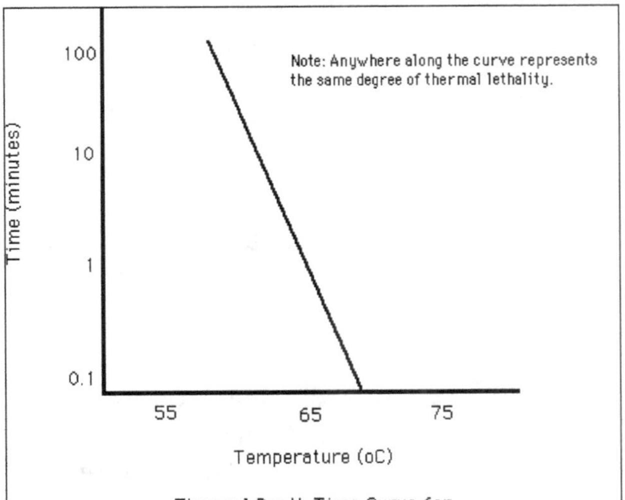

Thermal Death Time Curve for *Coxiella burnetti*, z = 4oC

Several parameters help us to do thermal calculations and define the rate of thermal lethality. The D value is a measure of the heat resistance of a microorganism. It is the time in minutes at a given temperature required to destroy 1 log cycle (90%) of the target microorganism. (Of course, in an actual process, all others that are less heat tolerant are destroyed to a greater extent). For example, a D value at 72 °C of 1 minute means that for each minute of processing at 72 °C the bacteria population of the target microorganism will be reduced by 90%. In the illustration below, the D value is 14 minutes (40-26) and would be representative of a process at 72 °C.

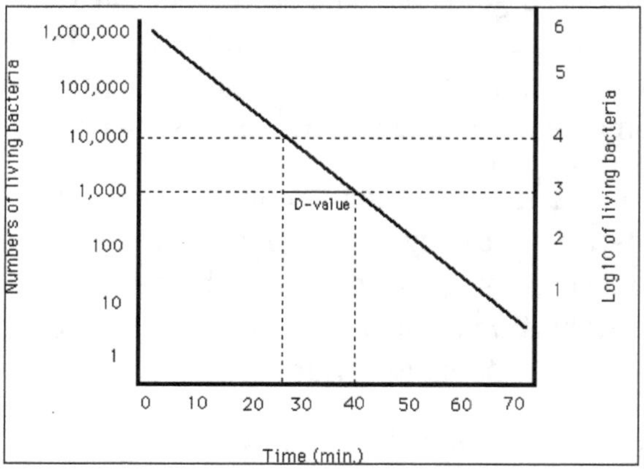

The Z value reflects the temperature dependence of the reaction. It is defined as the temperature change required to change the D value by a factor of 10. In the illustration below the Z value is 10 °C.

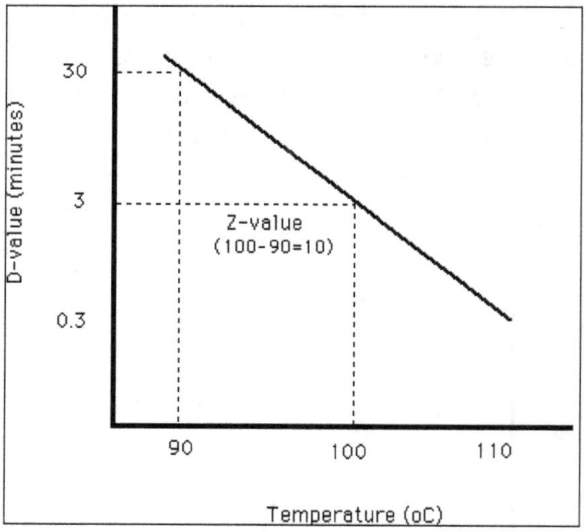

Reactions that have small Z values are highly temperature dependent, whereas those with large Z values require larger changes in temperature to reduce the time. A Z value of 10 °C is typical for a spore forming bacterium. Heat induced chemical changes have much larger Z values that microorganisms.

- Bacteria Z (°C) 5-10 D121 (min) 1-5.
- enzymes Z (°C) 30-40 D121 (min) 1-5.
- vitamins Z (°C) 20-25 D121 (min) 150-200.
- pigments Z (°C) 40-70 D121 (min) 15-50.

The figure below (which is schematic and not to scale) illustrates the relative changes in time temperature profiles for the destruction of microorganisms. Above and to the right of each line the microorganisms or quality factors would be destroyed, whereas below and to the left of each line, the microorganisms or quality factors would not be destroyed. Due to the differences in Z values, it is apparent that at higher temperatures for shorter times, a region exists (shaded area) where pathogens can be destroyed while vitamins can be maintained. The same holds true for other quality factors such as colour and flavour components. Thus in UHT milk processing, very high temperatures for very short times (e.g., 140 °C for 1-2 s) are favoured compared to a lower temperature longer time processes since it results in bacterial spore elimination with a lower loss of vitamins and better sensory quality.

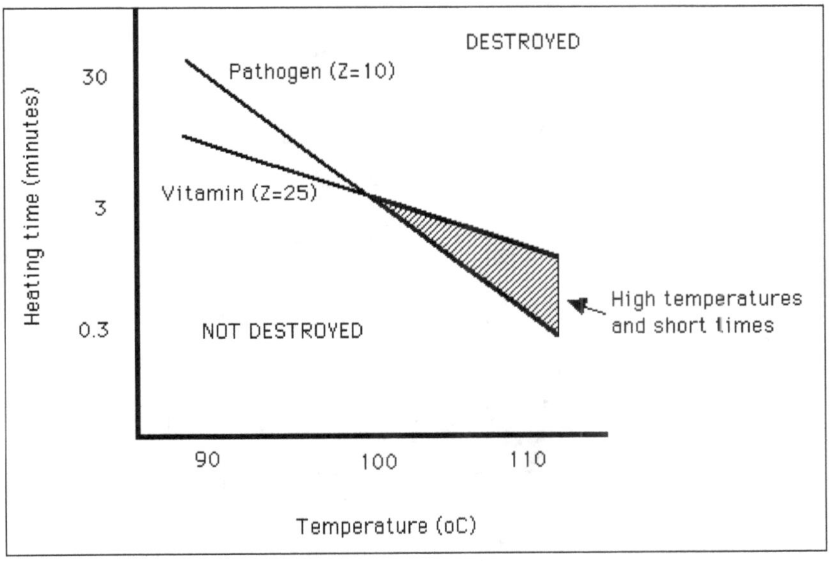

Alkaline phosphatase is a naturally-occurring enzyme in raw milk which has a similar Z value to heat-resistant pathogens. Since the direct estimation of pathogen numbers by microbial methods is expensive and time consuming, a simple test for phosphatase activity is routinely used. If activity is found, it is assumed that either the heat treatment was inadequate or that unpasteurized milk has contaminated the pasteurized product.

A working example of how to use D and Z values in pasteurization calculations:

Pooled raw milk at the processing plant has bacterial population of 4x10exp5/mL. It is to be processed at 79 °C for 21 seconds. The average D value at 65 °C for the mixed population is 7 min. The Z value is 7 °C. How many organisms will be left after pasteurization? What time would be required at 65 °C to accomplish the same degree of lethality?

Answer:

At 79 °C, the D value has been reduced by two log cycles from that at 65 °C since the Z value is 7 °C. Hence it is now 0.07 min. The milk is processed for 21/60=0.35 min, so that would accomplish 5 log cycle reductions to 4 organisms/mL. At 65 °C, you would need 35 minutes to accomplish a 5D reduction.

Methods of Pasteurization

Batch Method

The batch method uses a vat pasteurizer which consists of a jacketed vat surrounded by either circulating water with added steam or heating coils of hot water or direct steam.

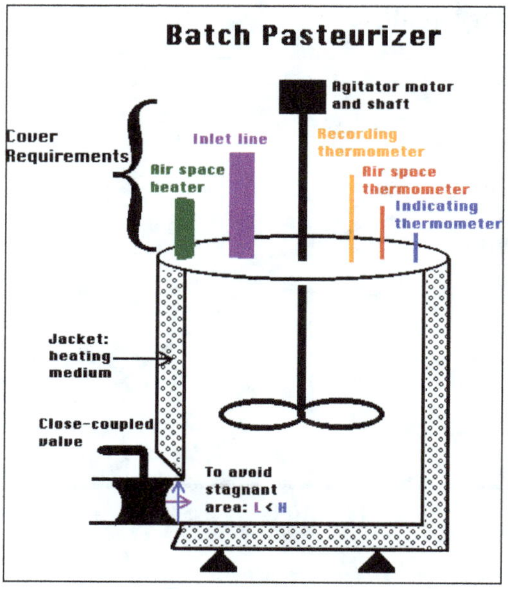

In the vat the milk is heated and held throughout the holding period while being agitated. The milk may be cooled in the vat or removed hot after the holding time is completed for every particle. As a modification, the milk may be partially heated in tubular or plate heater before entering the vat. This method has very little use for milk but some use for milk by-products (e.g. creams, chocolate) and special batches. The vat pasteurizer is used extensively in the ice cream industry as it allows for dissolution and blending of ingredients during the heating stage.

Continuous Method

Continuous process method has several advantages over the vat method, the most important being time and energy saving. For most continuous processing, a high temperature short time (HTST) pasteurizer is used. The heat treatment is accomplished using a plate heat exchanger. This piece of equipment consists of a stack of corrugated stainless steel plates clamped together in a frame. There are several flow patterns that can be used. Gaskets are used to define the boundaries of the channels and to prevent leakage. The heating medium can be vacuum steam or hot water.

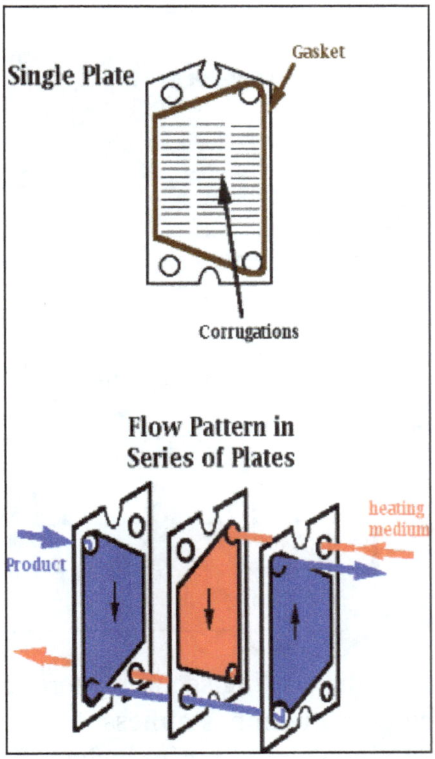

HTST Milk Flow

Cold raw milk at 4 °C in a constant level tank is drawn into the regenerator section of pasteurizer. Here it is warmed to approximately 57 °C - 68 °C by heat given up by hot pasteurized milk flowing in a counter current direction on the opposite side of thin, stainless steel plates. The raw milk, still under suction, passes through a positive displacement timing pump which delivers it under positive pressure through the rest of the HTST system.

The raw milk is forced through the heater section where hot water on opposite sides of the plates heat milk to a temperature of at least 72 °C. The milk, at pasteurization temperature and under pressure, flows through the holding tube where it is held for at least 16 sec. The maximum velocity is governed by the speed of the timing pump, diameter and length of the holding tube, and surface friction. After passing temperature sensors of an indicating thermometer and a recorder-controller at the end of the holding tube, milk passes into the flow diversion device (FDD). The FDD assumes a forward-flow position if the milk passes the recorder-controller at the preset cut-in temperature (>72 °C). The FDD remains in normal position which is in diverted-flow if milk has not achieved preset cut-in temperature. The improperly heated milk flows through the diverted flow line of the FDD back to the raw milk constant level tank. Properly heated milk flows through the forward flow part of the FDD to the pasteurized milk regenerator section where it gives up heat to the raw product and in turn is cooled to approximately 32 °C - 9 °C.

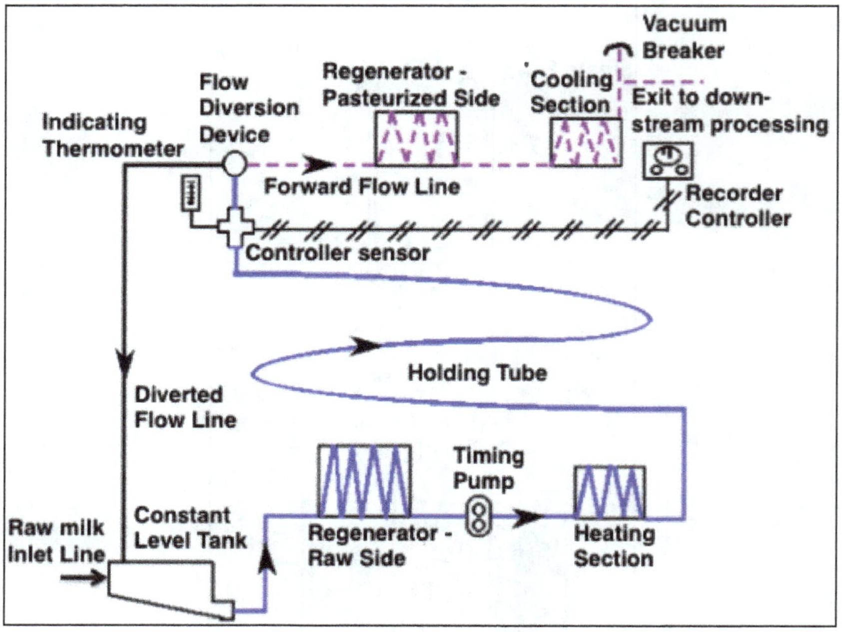

The warm milk passes through the cooling section where it is cooled to 4 °C or below by coolant on the opposite sides of the thin, stainless steel plates. The cold, pasteurized milk passes through a vacuum breaker at least 12 inches above the highest raw milk in the HTST system then on to a storage tank filler for packaging.

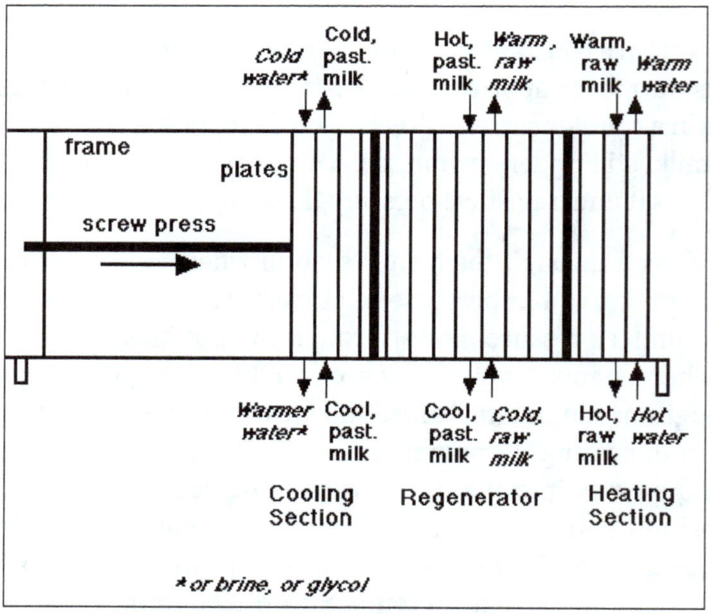

The diagram below illustrates HTST milk pasteurization equipment and flow diagram when incorporating a centrifugal booster pump (to help fill the regenerator on large systems) and pressure differential switch/back pressure controller on the regenerator (to maintain the pressure differential due to the action of the booster pump on the raw side). It also shows the position of the homogenizer when included in-line in the HTST system.

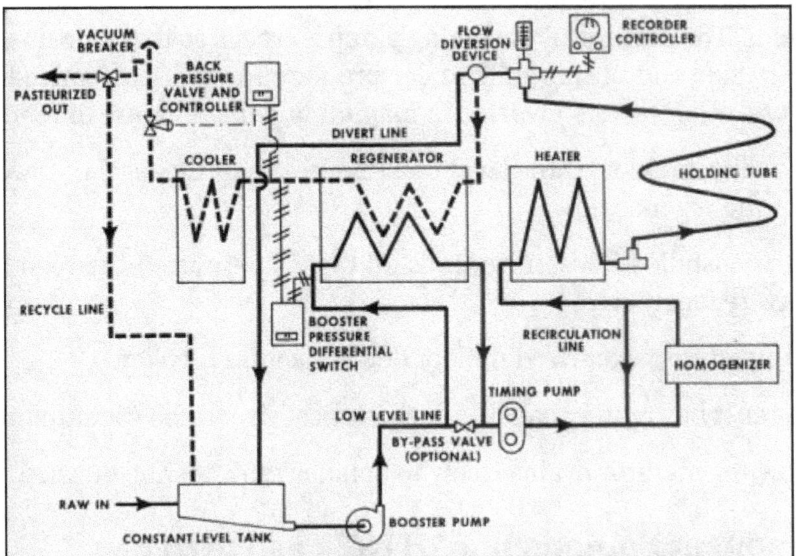

HTST PASTEURIZER WITH A BOOSTER PUMP, HOMOGENIZER OF LARGER CAPACITY THAN PASTEURIZER, FED BY A TIMING PUMP AND HAVING A REGENERATOR BACK PRESSURE CONTROL VALVE

Holding Time

When fluids move through a pipe, either of two distinct types of flow can be observed. The first is known as turbulent flow which occurs at high velocity and in which eddies are present moving in all directions and at all angles to the normal line of flow. The second type is streamline, or laminar flow which occurs at low velocities and shows no eddy currents. The Reynolds number, is used to predict whether laminar or turbulent flow will exist in a pipe:

- Re < 2100 laminar.
- Re > 4000 fully developed turbulent flow.

There is an impact of these flow patterns on holding time calculations and the assessment of proper holding tube lengths.

The holding time is determined by timing the interval for an added trace substance (salt) to pass through the holder. The time interval of the fastest particle of milk is desired. Thus the results found with water are converted to the milk flow time by formulation since a pump may not deliver the same amount of milk as it does water.

Pressure Differential

For continuous pasteurizing, it is important to maintain a higher pressure on the pasteurized side of the heat exchanger. By keeping the pasteurized milk at least 1 psi higher than raw milk in regenerator, it prevents contamination of pasteurized milk with raw milk in event that a pin-hole leak develops in thin stainless steel plates. This pressure differential is maintained using a timing pump in simple systems, and differential pressure controllers and back pressure flow regulators at the chilled pasteurization outlet in more complex systems. The position of the timing pump is crucial so that there is suction on the raw regenerator side and pushes milk under pressure through pasteurized regenerator. There are several other factors involved in maintaining the pressure differential:

- The balance tank overflow level must be less than the level of lowest milk passage in the regenerator.
- Properly installed booster pump is all that is permitted between balance tank and raw regenerator.
- No pump after pasteurized milk outlet to vacuum breaker.
- There must be greater than a 12 inch vertical rise to the vacuum breaker.
- The raw regenerator drains freely to balance tank at shut-down.

Basic Component Equipment of HTST Pasteurizer

Balance Tank

The balance, or constant level tank provides a constant supply of milk. It is equipped with a float valve assembly which controls the liquid level nearly constant ensuring uniform head pressure on the product leaving the tank. The overflow level must always be below the level of lowest milk passage in regenerator. It, therefore, helps to maintain a higher pressure on the pasteurized side of the heat exchanger. The balance tank also prevents air from entering the pasteurizer by placing the top of the outlet pipe lower than the lowest point in the tank and creating downward slopes of at least 2%. The balance tank provides a means for recirculation of diverted or pasteurized milk.

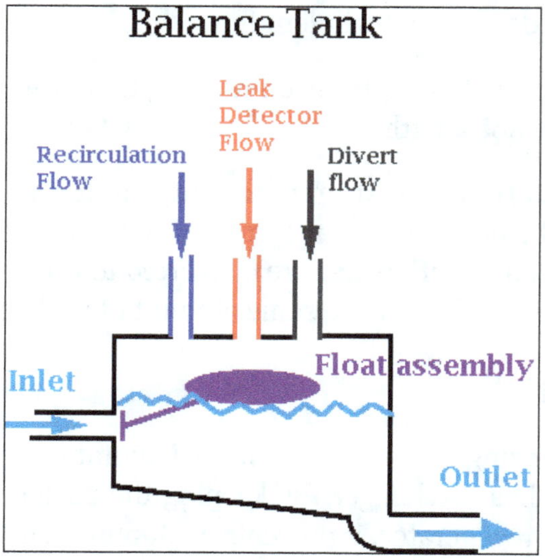

Regenerator

Heating and cooling energy can be saved by using a regenerator which utilizes the heat content of the pasteurized milk to warm the incoming cold milk. Its efficiency may be calculated as follows:

% regeneration = temp. increase due to regenerator/total temp. increase,

For example: Cold milk entering system at 4 °C, after regeneration at 65 °C, and final temperature of 72 °C would have an 89.7% regeneration:

$$\frac{65 - 4}{72 - 4} = 89.7$$

Timing Pump

The timing pump draws product through the raw regenerator and pushes milk under pressure through pasteurized regenerator. It governs the rate of flow through the holding tube. It must be a positive displacement pump equipped with variable speed drive that can be legally sealed at the maximum rate to give minimum holding time in holding tubes. It also must be interwired so it only operates when FDD is fully forward or fully diverted, and must be "fail-safe".A centrifugal pump with magnetic flow meter and controller may also be used.

Holding Tube

Must slope upwards 1/4"/ft. in direction of flow to eliminate air entrapment so nothing flows faster at air pocket restrictions.

Indicating Thermometer

The indicating thermometer is considered the most accurate temperature measurement. It is the official temperature to which thesafety thermal limit recorder (STLR) is adjusted. The probe should sit as close as possible to STLR probe and be located not greater than 18 inches upstream of the flow diversion device.

Recorder-controller (STLR)

The STLR records the temperature of the milk and the time of day. It monitors, controls and records the position of the flow diversion device (FDD) and supplies power to the FDD during forward flow. There are both pneumatic and electronic types of controllers. The operator is responsible for recording the date, shift, equipment, ID, product and amount, indicating thermometer temperature, cleansing cycles, cut in and cut out temperatures, any connects for unusual circumstances, and his/her signature.

Flow Diversion Device (FDD)

Also called the flow diversion valve (FDV), it is located at the downstream end of the upward sloping holding tube. It is essentially a 3-way valve, which, at temperatures greater than 72 °C,opens to forward flow. This step requires power. At temperatures less than 72 °C, the valve recloses to the normal position and diverts the milk back to the balance tank. It is important to note that the FDD operates on the measured temperature, not time, at the end of the holding period. There are two types of FDD:

- Single stem - An older valve system that has the disadvantage that it can't be cleaned in place.
- Dual stem - Consists of 2 valves in series for additional fail safe systems. This FDD can be cleaned in place and is more suited for automation.

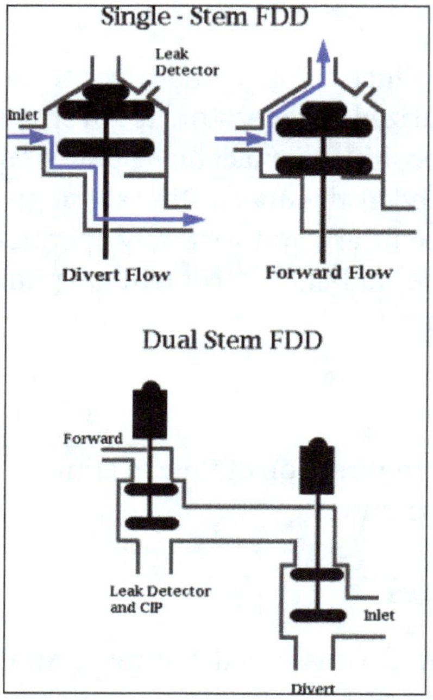

Vacuum Breaker

At the pasteurized product discharge is a vacuum breaker which breaks to atmospheric pressure. It must be located greater than 12 inches above the highest point of raw product in system. It ensures that nothing downstream is creating suction on the pasteurized side.

Auxiliary Equipment

Booster Pump

It is centrifugal "stuffing" pump which supplies raw milk to the raw regenerator for the balance tank. It must be used in conjunction with pressure differential controlling device and shall operate only when timing pump is operating, proper pressures are achieved in regenerator, and system is in forward flow.

Homogenizer

The homogenizer may be used as timing pump. It is a positive pressure pump; if not, then it cannot supplement flow. Free circulation from outlet to inlet is required and the speed of the homogenizer must be greater than the rate of flow of the timing pump.

Magnetic Flow Meter and Centrifugal Pump Arrangements

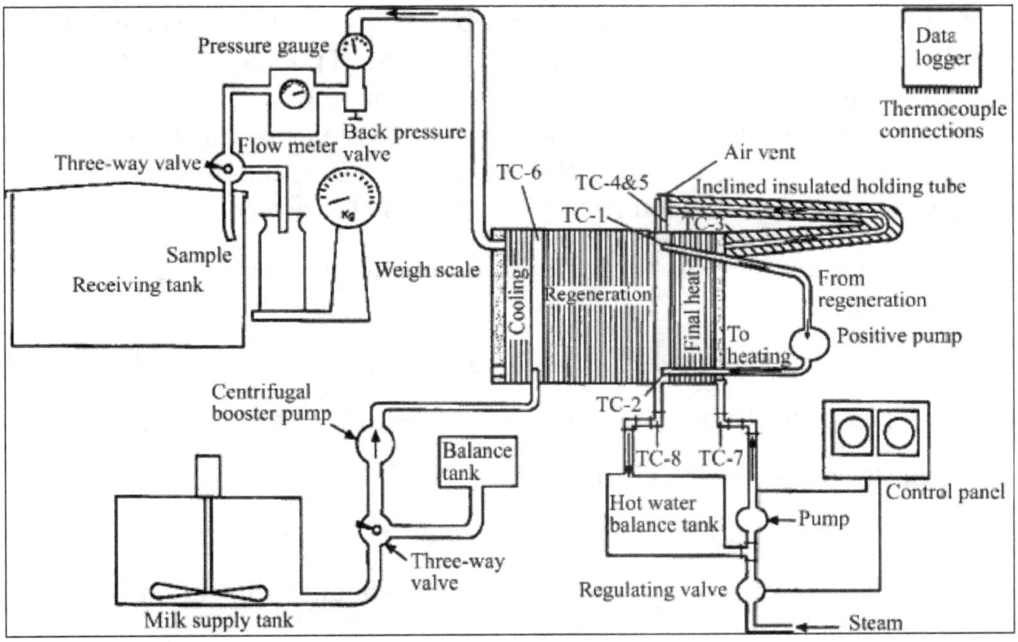

Magnetic flow meters can be used to measure the flow rate. It is essentially a short piece of tubing (approximately 25 cm long) surrounded by a housing, inside of which are located coils that generate a magnetic field. When milk passes through the magnetic field, it causes a voltage to be induced, and the generated signal is directly proportional to velocity. Application of the magnetic flow meter in the dairy industry has centered around its replacing the positive displacement timing pump as the metering device in HTST pasteurizing systems, where with certain products the timing pump rotors reportedly wear out in a relatively short period of time. In operation, the electrical signal is sent by the magnetic flow meter to the flow controller, which determines what the actual flow is compared to the flow rate set by the operator. Since the magnetic flow meter continuously senses flow rate, it will signal the electronic controller if the actual flow exceeds the set flow rate for any reason. If the flow rate is exceeded for any reason, the flow diversion device is put into diverted flow. A significant difference from the normal HTST system (with timing pump) comes into focus at this point. This system can be operated at a flow rate greater than (residence time less than) the legal limit. However, it will be in diverted flow and never in forward flow.

Another magnetic flow meter based system with an AC variable frequency motor control drive on a centrifugal pump is also possible in lieu of a positive displacement metering pump on a HTST pasteurizer. This system does not use a control valve but rather the signal from the magnetic flow meter is transmitted to the AC variable frequency control to vary the speed of the centrifugal pump. The pump, then controls the flow rate of product through the system and its holding time in the holding tube.

Automated Public Health Controllers

These systems are used for time and temperature control of HTST systems. There are concerns that with sequential control, the critical control points (CCP's) are not monitored all the time; if during the sequence it got held up, the CCP's would not be monitored. With operator control, changes can be made to the program which might affect CCP's; the system is not easily sealed. No computer program can be written completely error free in large systems; as complexity increases, so too do errors.

This gives rise to a need for specific regulations or computer controlled CCP's of public health significance:

- Dedicated computer - No other assignments, monitor all CCP's at least once/sec.
- Not under control of any other computer system or override system, i.e., network.
- Separate computer on each pasteurizer.
- I/O bus for outputs only, to other computers no inputs from other computers.
- On loss of power - Public health computers should revert to fail safe position (e.g. divert).
- Last state switches during power up must be fail safe position.
- Programs in ROM - Tapes/disks not acceptable.
- Inputs must be sealed, modem must be sealed, program sealed.
- No operator override switches.
- Proper calibration procedure during that printing - Public health computer must not leave public health control for >1 sec and upon return must complete 1 full cycle before returning to printing.
- FDV position must be monitored and temperature in holding tube recorded during change in FDV position.
- Download from ROM to RAM upon startup.

- Integrated with CIP computer which can be programmed e.g., FDV, booster pump controllable by CIP computer when in CIP made only.

Role of Biopreservation in Food Safety

Modern technologies in food processing and microbiological food safety standards have reduced but not eliminated the likelihood of food-related illness and product spoilage in industrialized countries. Food spoilage refers to the damage of the original nutritional value, texture, flavour of the food that eventually render food harmful to people and unsuitable to eat.

The increasing consumption of precooked food especially seafood, prone to temperature abuse, and the import of raw seafood from developing countries results in outbreak of food borne illness. One of the concerns in food industry is the contamination by pathogens, which are frequent cause of food borne diseases. In the USA, acute gastroenteritis affects 250 to 350 million people with more than 500 human deaths annually and approximately 22 to 30% of these cases are thought to be food borne diseases with the main foods implicated including meat, poultry, eggs, seafood and dairy products. Several bacterial pathogens including *Salmonella, Campylobacter jejuni, Escherichia coli 0157:H7, Listeria monocytogenes, Staphylococcus aureus* and *Clostridium botulinum* are found associated with such outbreaks.

In order to achieve improved food safety against such pathogens, food industry makes use of chemical preservatives or physical treatments (e.g. high temperatures). These preservation techniques have many drawbacks which includes the proven toxicity of the chemical preservatives (e.g. nitrites), the alteration of the organoleptic and nutritional properties of foods, and especially recent consumer demands for safe but minimally processed products without additives. To harmonize consumer demands with the necessary safety standards, traditional means of controlling microbial spoilage and safety hazards in foods are being replaced by combinations of innovative technologies that include biological antimicrobial systems such as lactic acid bacteria (LAB) and their metabolites.

The increasing demand for safe food has increased the interest in replacing chemical additives by natural products, without injuring the host or the environment. Biotechnology in the food-processing sector targets the selection, production and improvement of useful microorganisms and their products, as well as their technical application in food quality. The use of non-pathogenic microorganisms and their metabolites to improve microbiological safety and extend the shelf life of foods is defined as biopreservation. Antagonistic properties of LAB allied to their safe history of use in traditional fermented food products make them very attractive to be used as biopreservatives.

The use of non-pathogenic microorganisms and their metabolites to improve microbiological safety and extend the shelf life of foods is defined as biopreservation. Biopreservation refers to extended storage life and enhanced safety of foods using the natural microflora and their antibacterial products. It can be defined as the extension of shelf life and food safety by the use of natural or controlled microbiota and their antimicrobial compounds. One of the most common forms of food biopreservation is fermentation, a process based on the growth of microorganisms in foods, whether natural or added. It employs the breakdown of complex compounds, production of acids and alcohols, synthesis of Vitamin-B12, riboflavin and Vitamin-C precursor, ensures antifungal activity and improvement of organoleptic qualities such as, production of flavor and aroma compounds. In fish processing, biopreservation is achieved by adding antimicrobials or by increasing the acidity of the fish muscle. Efforts have concentrated on identification and development of protective bacterial cultures with antimicrobial effects against known pathogens and spoilage organisms. Following compounds such as organic acids, bacteriocins, diacetyl and acetaldehyde, enzymes, CO_2, hydrogen peroxide etc. are contributing to antimicrobial activity by Microbiota.

Bacteriocin

Bacteriocins are peptides or complex proteins biologically active with antimicrobial action against other bacteria, principally closely related species. They are produced by bacteria and are normally not termed antibiotics in order to avoid confusion and concern with therapeutic antibiotics, which can potentially illicit allergic reactions in humans and other medical problems. Bacteriocins differ from most therapeutic antibiotics in being proteinaceous agents that are rapidly digested by proteases in the human digestive tract. Since, bacteriocins are ribosomally synthesized; there exists a possibility of improving their characteristics to enhance their intensity and spectra of action. Colicine was the first bacteriocin, discovered in 1925 by André Gratia and his workgroup.

Bacteriocin production could be considered as an advantage for food and feed producers since, in sufficient amounts, these peptides can kill or inhibit pathogenic bacteria that compete for the same ecological niche or nutrient pool. This role is supported by the fact that many bacteriocins have a narrow host range, and is likely to be most effective against related bacteria with nutritive demands for the same scarce resources.

LAB Bacteriocins

A large number of new bacteriocins in lactic acid bacteria (LAB) have been characterized in recent years. Most of the new bacteriocins belong to the class II bacteriocins which are small (30–100 amino acids) heat-stable and commonly not post-translationally modified. While most bacteriocin producers synthesize only one bacteriocin, it has been shown that several LAB produce multiple bacteriocins (2–3 bacteriocins). The production of some class II bacteriocins (plantaricins of *Lactobacillus plantarum* C11 and sakacin P of *Lactobacillus sake*) have been shown to be transcriptionally regulated through a signal transduction system which consists of three components: An induction factor (IF), histidine protein kinase (HK) and a response regulator (RR).

Some bacteriocin-producing strains can be applied as protective cultures in a variety of food products and LAB bacteriocins possess many attractive characteristics that make them suitable candidates for use as food preservatives, such as:

- Protein nature, inactivation by proteolytic enzymes of gastrointestinal tract.
- Non-toxic to laboratory animals tested and generally non-immunogenic.
- Inactive against eukaryotic cells.
- Generally thermo-resistant (can maintain antimicrobial activity after pasteurization and sterilization).
- Broad bactericidal activity affecting most of the Gram-positive bacteria and some, damaged, Gram-negative bacteria including various pathogens such as *Listeria monocytogenes, Bacillus cereus, Staphylococcus aureus* and *Salmonella.*

Genetic determinants generally located in plasmid, which facilitates genetic manipulation to increase the variety of natural peptide analogues with desirable characteristics.

In general, the following features should be considered when selecting bacteriocin-producing strains for food applications:

- The producing strain should preferably have GRAS (generally regarded as safe) status.
- Depending on the application, the bacteriocin should have a broad spectrum of inhibition that includes pathogens or else high specific activity.
- Thermostability.
- Beneficial effects and improved safety.
- No adverse effect on quality and flavour.

Biopreservation with Bacteriocin: Field of Application

Before bacteriocin can be applied in foods their cytolytic abilities should be assessed in detail. This is a very important issue since recently a cytolysin produced by *E. faecalis* was described that possesses both haemolytic and bacteriocin activities Recombinant DNA technology is currently applied, to enhance production, to transfer bacteriocin genes to other species and for mutation and selection of bacteriocin variants with increased and broad activity spectra. Continued study of the physical and chemical properties, mode of action and structure-function relationships of bacteriocins is necessary if their potential in food preservation is to be exploited. Further research into the synergistic reactions of these compounds and other natural preservatives, in combination with advanced technologies could result in replacement of chemical preservatives, or could allow less severe processing (eg. heat) treatments, while still maintaining adequate microbiological safety and quality in foods.

Classification

On the basis of structure and mode of action bacteriocins are divided in 4 major groups.

Class I

- It is termed lantibiotics, constitue a group of small peptides that are characterized by their content of several unusual amino acids.
- They are small peptides that are differentiated from other bacteriocins by their content in dehydroamino acids and thioether amino acids. They include nisin, discovered in 1928, lacticin 481 of *L. lactis*, citolysin of *E. faecalis*, and lacticin 3147 of *L. lactis*, among others.

Class II

- These are small, nonmodified, heat stable peptides.
- In general, they have an amphiphilic helical structure, which allows them to insert into the membrane of the target cell, leading to depolarisation and death.
- They comprises the (<10 kDa) thermostable non-lantibiotic linear peptides. They are divided into three subclasses on the basis of either a distinctive N-terminal sequence, the pediocin- like bacteriocins (class II.1) (e.g. pediocin PA- 1/ AcH produced by *Pediococcus*, the lack of leader peptide (class II.2) (e.g. enterocin EJ97 by *E. faecalis*, or neither of the above traits (class II.3)(e.g. enterocin L50A by *E. faecalis*.

Class III

- It is formerly termed bacteriolysins, large (> 30 KDa), heat-labile protein bacteriocins, such as helveticin J of *L. helveticus* and bacteriocin Bc-48 of *E. faecali*.
- It can function directly on the cell wall of Gram- positive targets, leading to death and lysis of the target cell.

Class IV

- It is presently reserved for cyclic bacteriocins composed not only from protein (also lipid or cidrate).

Factors Inhibiting Bacteriocin Production

- Inadequate physical conditions and chemical composition of food (pH, temperature, nutrients, etc.);
- Spontaneous loss in production capacity;
- Inactivation by phage of the producing strain;
- Antagonism effect of other microorganisms in foods.

The Effectiveness of Bacteriocin Activity in Food is Negatively affected by:

- Resistance development of pathogens to the bacteriocin;
- Inadequate environmental conditions for the biological activity;
- Higher retention of the bacteriocin molecules by food system components (e.g. fat);
- Inactivation by other additives; slower diffusion and solubility and irregular distribution of bacteriocin molecules in the meat matrix.

Several factors, such as the presence of salts, other food ingredients, poor solubility and the uneven distribution of the bacteriocin, have all shown to effect the efficacy of bacteriocins in food.

Bacteriocins of Various Gram Positive Bacteria

Although mast research has focused on antimicrobial agents produced by lactic acid bacteria, many bacteriocins of other Gram positive bacteria have been isolated and documented. One such area of interest is the use of these bacteriocins to control the growth of undesirable microorganisms, particularly those of public health concern, e.g., *C. botulinum* and *L. monocytogenes*.

Bacteriocin production has been documented for a variety of Gram positive bacteria, including *Staphylococcus*, *Clostridium*, and *Bacillus spp*. Although the use of these bacteriocins may be precluded from foods because the producer strain may be a pathogen, recent developments in genetic engineering techniques have made the transfer of genes encoding for bacteriocin production from bath Gram positive and Gram negative bacteria to food grade microorganisms possible.

Bacillus as Biopreservative

Bacteriocin production in *Bacillus spp*. has been studied over the past few decades and several reports describe the production, isolation and characterization of bacteriocins from these species which include Subtilin from *B. subtilis*, Megacin from *B. megaterium* and Thermacin from *B. stearothermophilus*.

Subtilin, a cationic peptide produced by *B. subtilis*, having molecular mass of 3317 Da.; is a member of the group of bacteriocins known as the lantibiotics. The structure of subtilin was determined by Gross.

Bacteriocins produced by *Bacillus spp.* could be alternatives to those produced by lactic acid bacteria for several reasons:

- *Bacillus spp.*, like lactic acid bacteria, have been used for hundreds of years in making food and various enzymes from *Bacillus* have been used intensively in food processing worldwide. No adverse effects have been demonstrated in humans from consuming foods made from *Bacillus spp.*, and their products. Bacteriocins from these microorganisms would be safe for humans and would be no more of a risk than lactic acid bacteria.
- *Bacillus spp.* have an antimicrobial action against Gram positive and Gram negative bacteria, as well as fungi, and therefore have a greater antimicrobial spectra than lactic acid bacteria and their bacteriocins.
- The metabolic diversities of *Bacillus spp.* may result in bacteriocins with various properties such as inhibitory activity at alkaline, acidic condition or after thermal processing and would be suitable for food processing.
- The physiology/genetics of *Bacillus* are well understood, second only to those of *Escherichia coli*. Molecular biological techniques would provide safe/reliable tools for producing bacteriocins for the food industry.

Biopreservation of Seafood Products

Although bacteriocins are produced by many Gram-positive and Gram-negative species, those produced by LAB and now a days, *Bacillus spp.* are of particular interest to the food industry, since these bacteria have generally been regarded as safe. Among the lactic acid bacteria, a high diversity of bacteriocins is produced and several have been patented for their applications in foods. To date, the only commercially produced bacteriocins are the group of nisins produced by *Lactoccocus lactis*, and pediocin PA-1, produced by *Pediococcus acidilactic.*

Bacteriocins produced by lactic acid bacteria (LAB) have received particular attention in recent years due to their potential application in food industry as natural preservatives. Biopreservation refers to the use of antagonistic microorganisms or their metabolic products to inhibit or destroy undesired microorganisms in foods to enhance food safety and extend shelf life. Three approaches are commonly used in the application of bacteriocins for biopreservation of foods:

- Inoculation of food with LAB that produce bacteriocin in the products. The ability of the LAB to grow and produce bacteriocin in the products is crucial for its successful use.

- Addition of purified or semi-purified bacteriocins as food preservatives.
- Use of a product previously fermented with a bacteriocin producing strain as an ingredient in food processing.

The effectiveness of bacteriocins and protective cultures to control growth of L. monocytogenes in vacuum-packed cold smoked salmon has been demonstrated by several researchers. Katla examined the inhibitory effect of sakacin P and L. sake cultures (sakacin P producer) against L. monocytogenes in cold-smoked salmon. Nilsson showed that a non- bacteriocin-producing strain of C. piscicola was as effective as a bacteriocin-producing strain of C. piscicola in the inhibition of L. monocytogenes in vacuum-packed cold-smoked salmon. They suggested that the growth inhibition of L. monocytogenes was probably due to the competitive growth of C. piscicola that resulted in depletion of essential nutrients. The inhibitory effect of nisin in combination with carbon dioxide and low temperature on the survival of L. monocytogenes in cold-smoked salmon has also been investigated. The effectiveness of nisin Z, carnocin UI49, and a preparation of crude bavaricin A on shelflife extension of brined shrimp was evaluated by Einarsson and Lauzon. In a study using vacuum-packed cold-smoked rainbow trout, Niskänen examined the inhibition of L. monocytogenes and mesophilic aerobic bacteria by nisin, sodium lactate or their combination.

References

- Safe-food-handling-1150: dayjob.com, Retrieved 10 January 2018
- Everything-you-need-to-know-about-iso-22000: foodsafety.com.au, Retrieved 17 July 2018
- Hurdle-Technology-in-Food-Preservation-301930970: researchgate.net, Retrieved 20 March 2018
- Food-irradiation-what-you-need-know, buy-store-serve-safe-food: fda.gov, Retrieved 17 June 2018
- Role-of-Biopreservation-in-Improving-Food-Safety-and-Storage-317078458: researchgate.net, Retrieved 28 April 2018

3

Preservation of Food

The food can be preserved by using many preservation techniques such as low-temperature preservation, thermal and non-thermal processing of food, controlling water activity, fermentation, food additive, chemical preservation, etc. This chapter has been carefully written to provide an easy understanding of these techniques associated with preservation of food.

Food Preservation

Food preservation can be defined as the science that deals with the process of prevention of decay or spoilage of food thus, allowing it to be stored in a fit condition for future use.

In other words, food preservation is the process in which the perishable food materials are given a suitable physical or chemical treatment to prevent their wastage or spoilage and to retain their nutritive value for long periods.

Microorganisms are responsible for causing or accelerating spoilage but also in some cases, use of benign bacteria and yeasts add specific qualities and preserve food e.g. cheese, wine. Spoilage results in loss of quality, edibility or nutritive value of food. Preservation usually involves the process of treating and handling food to stop or greatly slow down spoilage as well as maintaining or creating nutritional value, texture and flavour of food.

Preservation mainly prevents the growth of bacteria, fungi, and other microorganisms. It also includes processes to check the oxidation of fats which causes "Rancidification"rancidity and to inhibit natural ageing and discolouration due to enzymatic browning. In some cases, preservation also includes sealing of food materials after heat treatment to avoid recontamination with microbes, such as canning. Methods like drying, allow food to be stored without any special control for long periods.

A number of preservation methods include drying, spray drying, freeze drying, freezing, vacuum-packing, canning, food irradiation, and adding preservatives or inert "Gas"gases such as carbon dioxide. Other methods that not only help to preserve food, but also add flavour, include pickling, salting, smoking, preserving in sugar syrup and curing.

Importance of Food Preservation

Food production and supply does not always tally with the demand or meets of the people. In some places there is surplus production of a food product, whereas in some other place there is inadequate supply. Even foods are perishable and semi-perishable like juicy fruits, vegetables, mangoes, tomato, papaya and many more, which very quickly gets spoilt. It is therefore important to improve and expand facilities for storage and preservation of food. Food preservation helps in:

- Increasing the self-life of foods thus increasing the supply. So many perishable foods can be preserved for a long time.
- Making the seasonal food available throughout the year.
- Adding variety to the diet.
- Saving time by reducing preparation time and energy, as the food has already been partially processed.
- Stabilising prices of food, as there is less scope of shortage of supply to demand.
- Decreasing wastage of food by preventing decay or spoilage of food.
- Improving the nutrition of the population. Preserved foods help people to bring a variety in the diet, thereby decreasing nutritional inadequacies.

Low-temperature Preservation

Use of low-temperature reduces the microbial activity and enzyme activity thus prolongs shelf life of foods. Two different temperatures are employed in low temperature namely chilling temperature and freezer temperature.

Chilling Technique

Chilled foods are those foods stored at temperatures near, but above their freezing point, typically 0-5 °C. This commodity area has shown a massive increase in recent years as traditional chilled products such as fresh meat and fish and dairy products have been joined by a huge variety of new products including complete meals, prepared and delicatessen salads, dairy desserts and many others.

Three main factors have contributed to this development:

- The food manufacturers' objective of increasing added value to their products;
- Consumer demand for fresh foods while at the same time requiring the convenience of only occasional shopping and ease of preparation; and
- The availability of an efficient cold chain – the organization and infrastructure which allows low temperatures to be maintained throughout the food chain from manufacture/harvest to consumption.

Chill storage can change both the nature of spoilage and the rate at which it occurs. There may be qualitative changes in spoilage characteristics as low temperatures exert a selective effect preventing the growth of mesophiles and leading to a microflora dominated by psychrotrophs.

This can be seen in the case of raw milk which in the days of milk churns and roadside collection had a spoilage microflora comprised largely of mesophiliclactococci which would sour the milk.

Nowadays in the UK, milk is chilled almost immediately it leaves the cow so that psychrotrophic Gram-negative rods predominate and produce an entirely different type of spoilage. Low temperatures can also cause physiological changes in microorganisms that modify or exacerbate spoilage characteristics.

Two such examples are the increased production of phenazine and carotenoid pigments in some organisms at low temperatures and the stimulation of extracellular polysaccharide production in Leuconostoc spp. and some other lactic acid bacteria.

In most cases, such changes probably represent a disturbance of metabolism due to the differing thermal coefficients and activation energies of the numerous chemical reactions that comprise microbial metabolism.

Though psychrotrophs can grow in chilled foods they do so only relatively slowly so that the onset of spoilage is delayed. In this respect temperature changes within the chill temperature range can have pronounced effects.

For example, the generation time for one pseudomonad isolated from fish was 6.7 hours at 5 °C compared with '26.6 hours at 0 °C. Where this organism is an important contributor to spoilage, small changes of temperature will have major implications for shelf-life.

The keeping time of haddock and cod fillets has been found to double if the storage temperature is decreased from 2.8 °C to — 0.3 °C. Mathematical modelling techniques of the sort can be useful in predicting the effect of temperature fluctuations on shelf-life, but, as a general rule, storage temperature should be as low, and as tightly controlled, as possible.

The ability of organisms to grow at low temperatures appears to be particularly associated with the composition and architecture of the plasma membrane. As the temperature is lowered, the plasma membrane undergoes a phase transition from a liquid crystalline state- to a rigid gel in which solute transport is severely limited.

The temperature of this transition is lower in psychrotrophs and psychrophiles largely as a result of higher levels of unsaturated and short chain fatty acids in their membrane lipids. If some organisms are allowed to adapt to growth at lower temperatures they increase the proportion of these components in their membranes.

There seems to be no taxonomic restriction on psychrotrophic organisms which can be found in the yeasts, moulds, Gram-negative and Gram-positive bacteria. One feature they share is that in addition to their ability to grow at low- temperatures, they are inactivated at moderate temperatures.

A number of reasons for this marked heat sensitivity have been put forward including the possibility of excessive membrane fluidity at higher temperatures. Low thermal stability of key enzymes and other functional proteins appears to be an important factor, although thermo-stable extracellular lipases and proteases produced by psychrotrophic pseudomonads can be a problem in the dairy industry.

Though mesophiles cannot grow at chill temperatures, they are not necessarily killed. Chilling will produce a phenomenon known as cold shock which causes death and injury in a proportion of the population but its effects are not predictable in the same way as heat processing.

The extent of cold shock depends on a number of factors such as the organism (Gram-negatives appear more susceptible than Gram- positives), its phase of growth (exponential-phase cells are more susceptible than stationary phase cells), the temperature differential and the rate of cooling (in both cases the larger it is, the greater the damage), and the growth medium cells grown in complex media are more resistant).

The principal mechanism of cold shock appears to be damage to membranes caused by phase changes in the membrane lipids which create hydroppores through which cytoplasmic contents can leak out. An increase in single-strand breaks in DNA has also been noted as well as the synthesis of specific cold-shock proteins.

Since chilling is not a bactericidal process, the use of good microbiological quality raw materials and hygienic handling are key requirements for the production of safe chill foods. Mesophiles that survive cooling, albeit in an injured state, can persist in the food for extended periods and may recover and resume growth should conditions later become favourable.

Thus chilling will prevent an increase in the risk from mesophilic pathogens, but will not assure its elimination. There are however pathogens that will continue to grow at some chill temperatures and the key role of chilling in the modern food industry has focused particular attention on these.

Risks posed by these organisms, may increase with duration of storage but this process is likely to be slow and dependent on the precise storage temperature and composition of the food.

Some foods are not amenable to chill storage as they suffer from cold injury where the low temperature results in tissue breakdown which leads to visual defects and accelerated microbiological deterioration. Tropical fruits are particularly susceptible to this form of damage.

Freezing Technique

Freezing is the most successful technique for long term preservation of food since nutrient content is largely retained and the product resembles the fresh material more closely than in appertized foods.

Foods begin to freeze somewhere in the range — 0.5 to — 3 °C, the freezing point being lower than that of pure water due to the solutes present. As water is converted to ice during freezing, the concentration of solutes in the unfrozen water increases, decreasing its freezing point still further so that even at very low temperatures, e.g. — 60 °C, some water will remain unfrozen.

The temperatures used in frozen storage are generally less than — 18 °C. At these temperatures no microbial growth is possible, although residual microbial or endogenous enzyme activity such as lipases can persist and eventually spoil a product.

This is reduced in the case of fruits and vegetables by blanching before freezing to inactivate endogenous polyphenol oxidases which would otherwise cause the product to dis-colour during storage.

Freezer burn is another non-microbiological quality defect that may arise in frozen foods, where surface discolouration occurs due to sublimation of water from the product and its transfer to colder surfaces in the freezer. This can be prevented by wrapping products in a water-impermeable material or by glazing with n layer of ice.

Low temperature is not the only inhibitory factor operating in frozen foods; they also have a low water activity produced by removal of water in the form of ice. Table describes the effect of temperature on water activity. As far as microbiological quality is concerned, this effect is only significant when frozen foods are stored at temperatures where microbial growth is possible (above — 10 °C).

Table: Effect of freezing on the water activity of pure water/ice.

Temperature (°C)	a_w
0	1
-5	0.953
-10	0.907
-15	0.864
-20	0.823
-40	0.68

In this situation, the organisms that grow on a product are not those normally associated with its spoilage at chill temperatures but yeasts and moulds that are both psychrotrophic and tolerant of reduced water activity.

Thus meat and poultry stored at – 5 to – 10 °C may slowly develop surface defects such as black spots due to the growth of the mould Cladosporium herbarum, white spots caused by Sporotrichum carnis or the feathery growth of Thamnidium elegans.

Micro-organisms are affected by each phase of the freezing process. In cooling down to the temperature at which freezing begins, a proportion of the population will be subject to cold shock.

At the freezing temperature, further death and injury occur as the cooling curve levels out as latent heat is removed and the product begins to freeze. Initially ice forms mainly extracellularly, intracellular ice formation being favoured by more rapid cooling.

This may mechanically damage cells and the high extracellular osmotic pressures generated will dehydrate them. Changes in the ionic strength and pH of the water phase as a result of freezing will also disrupt the structure and function of numerous cell components and macromolecules which depend on these factors for their stability.

Cooling down to the storage temperature will prevent any further microbial growth once the temperature has dropped below – 10 °C. Finally, during storage there will be an initial decrease in viable numbers followed by slow decline over time. The lower the storage temperature, the slower the death rate.

As with chilling, freezing will not render an unsafe product safe – its microbial lethality is limited and preformed toxins will persist. Frozen chickens are, after all, an important source of Salmonella.

Survival rates after freezing will depend on the precise conditions of freezing, the nature of the food material and the composition of its microflora, but have been variously recorded as between 5 and 70%. Bacterial spores are virtually unaffected by freezing, most vegetative Gram-positive bacteria are relatively resistant and Gram-negatives show the greatest sensitivity.

While frozen storage does reliably inactivate higher organisms such as pathogenic protozoa and parasitic worms, food materials often act as cryoprotectants for bacteria so that bacterial pathogens may survive for long periods in the frozen state. In one extreme example Salmonella has been successfully isolated from ice cream stored at – 23 °C for 7 years.

The extent of microbial death is also determined by the rate of cooling.

Maximum lethality is seen with slow cooling where, although there is little or no cold shock experienced by the organisms, exposure to high solute concentrations is prolonged. Survival is greater with rapid freezing where exposure to these conditions is minimized. Food freezing processes are not designed however to maximize microbial lethality but to minimize loss of product quality.

Formation of large ice crystals and prolonged exposure to high osmotic pressure solutions during slow cooling also damage cells of the food material itself causing greater drip loss and textural deterioration on thawing, so fast freezing in which the product is at storage temperature within half an hour is the method of choice commercially.

The rate of freezing in domestic freezers is much slower so, although microbial lethality may be greater, so too is product quality loss.

Thawing of frozen foods is a slower process than freezing. Even with moderate size material the outside of the product will be at the thawing temperature some time before the interior. So with high thawing temperature, mesophiles may be growing on the surface of a product while the interior is still frozen. Slow thawing at lower temperature is generally preferred.

It does have some lethal effect as microbial cells experience adverse conditions in the 0 to — 10 °C range for longer, but it will also allow psychrotrophs to grow. Provided the product is not subject to contamination after thawing, the microflora that develops will differ from that on the fresh material due to the selective lethal effect of freezing.

Lactic acid bacteria are often responsible for the spoilage of defrosted vegetables whereas they generally comprise only about 1% of the microflora on fresh chilled produce which is predominantly Gram-negative.

Freezing and defrosting may make some foods more susceptible to microbiological attack due to destruction of antimicrobial barriers in the product and condensation, but defrosted foods do not spoil more rapidly than those that have not been frozen. Injunctions against refreezing defrosted products are motivated by the loss of textural and other qualities rather than any microbiological risk that is posed.

Thermal Processing of Food

There are two main temperature categories employed in thermal processing: Pasteurization and Sterilisation. The basic purpose for the thermal processing of foods is to reduce or destroy microbial activity, reduce or destroy enzyme activity and to produce physical or chemical changes to make the food meet a certain quality standard. e.g. gelatenization of starch & denaturation of proteins to produce edible food. There are a number of types of heat processing employed by the food industry.

Mild processes	Blanching Pasteurisation
More severe processes	Canning Baking Roasting Frying

Blanching

The primary purpose of blanching is to destroy enzyme activity in fruit and vegetables. It is not intended as a sole method of preservation, but as a pretreatment prior to freezing, drying and canning. Other functions of blanching include:

- Reducing surface microbial contamination.
- Softening vegetable tissues to facilitate filling into containers.
- Removing air from intercellular spaces prior to canning.

Blanching and Enzyme Inactivation

Freezing and dehydration are insuffcient to inactivate enzymes and therefore blanching can be employed. Canning conditions may allow suffcient time for enzyme activity. Enzymes are proteins which are denatured at high temperatures and lose their activity. Enzymes which cause loss of quality include Lipoxygenase, Polyphenoloxidase, Polygaacturonase and Chlorophyllase. Heat resistant enzymes include Catalase and Peroxidase.

Methods of Blanching

Blanching is carried out at up to 100 °C using hot water or steam at or near atmospheric pressure.

Some use of fluidised bed blanchers, utilising a mixture of air and steam, has been reported. Advantages include faster, more uniform heating, good mixing of the product, reduction in effluent, shorter processing time and hence reduced loss of soluble and heat sensitive components.

There is also some use of microwaves for blanching. Advantages include rapid heating and less loss of water soluble components. Disadvantages include high capital costs and potential diffculties in uniformity of heating.

Steam Blanchers

This is the preferred method for foods with large cut surface areas as lower leaching losses. Normally food material carried on a mesh belt or rotatory cylinder through a steam atmosphere, residence time controlled by speed of the conveyor or rotation. Often poor uniformity of heating in the multiple layers of food, so attaining the required time-temperature at the centre results in overheating of outside layers.

Individual Quick Blanching (IQB) involves a first stage in which a single layer of the food is heated to suffcient temperature to inactivate enzymes and a second stage in which a deep bed of the product is held for suffcient time to allow the temperature at the centre of each piece to increase to that needed for inactivation.

The reduced heating time (e.g. for 10 mm diced carrot, 25 s heating and 50 s holding compared with 3 minutes conventional blanching) results in higher energy effcienciess For small products es gs peas, sliced or diced carrots), mass of produce blanched per kg steam increases from 0.5 kg for conventional steam blanchers to 6-7 kg for IQB.

Hot Water Blanchers

Includes various designs which hold the food in hot water (70 to 100 °C) for a specified time, then moves it to a dewatering/cooling section. In blanchers of this type the food enters a slowly rotating drum, partially submerged in the hot water. It is carried along by internal flights, residence time being controlled by the speed of rotation.

Pipe blanchers consist of insulated tubes through which hot water is circulated. Food is metered into the stream, residence time being controlled by the length of the pipe and velocity of the water.

The blancher-cooker has three sections, a preheating stage, a blanching stage, and a cooling stage. As the food remains on a single belt throughout the process, it is less likely to be physically damaged. With the heat recovery incorporated in the system, 16 to 20 kg of product can be blanched for every kg of steam, compared with 0.25 to 0.5 kg per kg stream in the conventional hot water blanchers.

Testing of the Effectiveness of Blanching

Over blanching causes quality loss due to overheating while under blanching causes quality loss due to increased enzyme activity because enzymes activated and substrates released by heat. The Peroxidase test in vegetables is used to detect enzyme inactivation. This enzyme is not in itself implicated in degradation, but is relatively heat resistant and easily detected. It consists of adding guaiacol solution and hydrogen peroxide solution and observing the development of a brown colour indicating peroxidase activity.

Complete inactivation is not always essential – green beans, peas and carrots with some residual peroxidase activity have shown adequate storage quality at -20 °C through with other vegetable (e.g. Brussels sprouts) zero peroxidase activity is essential.

Pasteurization

Purpose of Pasteurization

Pasteurization is a relatively mild heat treatment in which food is heated to <100 °C. It is widely used throughout the food industry and is frequently employed as a CCP in various HACCP plans. As a unit operation in food processing it can be used to destroy enzymes and relatively heat sensitive micro-organisms (e.g. non spore forming bacteria, yeast and moulds). In this regard is it used to extend shelf life by several days e.g. milk or months e.g. bottled fruit.

The severity of treatment and resulting extension of shelf life is determined mostly by pH of the food. In low acid foods (pH <4.5), the main purpose is destruction of pathogenic bacteria, while below pH 4.5 the destruction of spoilage microorganisms or enzyme deactivation is usually more important. The extent of heat treatment required is determined by the D value (Decimal reduction time or time to reduce numbers by a factor of 10 or 90% of the initial load) of most heat resistant enzyme or micro-organism which may be present. In terms of checking the effectiveness of the process, alkaline phosphatase is a naturally occurring enzyme in raw milk with a similar D value to heat-resistant pathogens and so is routinely used as an indicator of adequate pasteurisation. If phosphatase activity is found, it is assumed that pasteurisation is inadequate.

Pasteurization is normally used for the destruction of all disease causing organisms (e.g. pasteurization of milk) or the destruction or reduction in the number of spoilage organisms in certain foods e.g. vinegar.

Table: Milk pasteurizing temperatures.

Temperature	Time
63 °C	For 30 min (low temperature long time LTLT).
72 °C	For 15 sec (primary high temperature short time, HTST method).
89 °C	For 1.0 sec.
90 °C	For 0.5 sec.
94 °C	For 0.1 sec.
100 °C	For 0.01 sec.

These temperatures are equivalent and are suffcient to destroy the most heat sensitive of the non-spore-forming pathogenic organisms. Milk pasteurization temperatures are also suffcient to destroy all yeasts, moulds, gram negative bacteria and many gram positive. The two groups of micro-organisms that survive pasteurisation temperatures used in milk are:

Thermoduric: Organisms that can survive exposure to relatively high temperatures but do not necessarily grow at these temperatures e.g. Streptococcus and Lactobacillus.

Thermophilic: Organisms that not only survive relatively high temperatures but require high temperatures for their growth.

Method for Pasteurizing

There are number of basic methods of pasteurization widely used in the industry.

Batch (Holding) Method

In this method every particle (e.g. milk) must be heated to at least 63 °C and held for at least 30 minutes, however this is not used commercially these days.

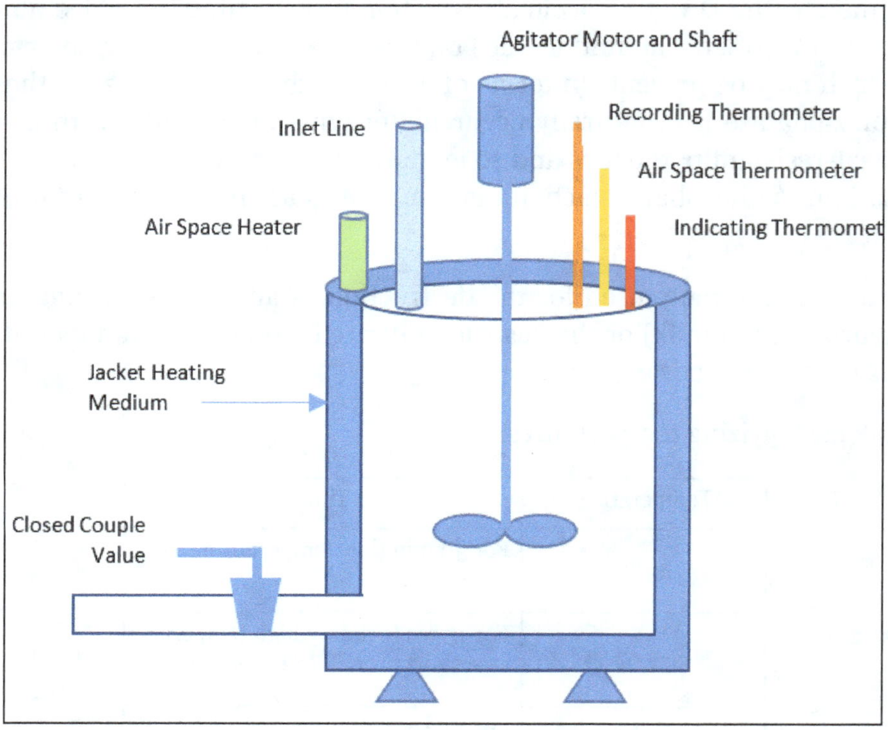

Batch Pasteurizer.

High-temperature Short-time (HTST)

In this method the heating of every particle of milk to at least 72 °C and holding for at least 15 seconds. Carried out as a continuous process. Ultra Heat Treatment (UHT) a sterilisation treatment, can also be performed using higher temperatures and shorter times e.g. 1 s at 135 °C.

Typical Equipment employed for this method includes:

- Plate heat exchanger (PHE).
- Holding tube – Sized to ensure the correct treatment time is achieved.
- Holding tanks – For storage of the raw and pasteurised milk.
- Balance tank – To assist in maintaining full flow, and to take returned milk if temperature not achieved.
- Control and monitoring system – To record temperature and to divert flow back to the balance tank if correct temperature is not achieved.

Pasteurization of Packaged Foods

Some liquid foods (e.g. beer and fruit juices) are pasteurized after filling into containers. Hot water is normally used if the food is packaged into glass, to reduce the risk of breakage due to thermal shock. Maximum temperature between the container and the liquid are 20 °C for heating and 10 °C for cooling. Metal and plastic containers may be pasteurized using steam-air mixtures or hot water. Pasteurisers may be batch or continuous. A simple batch type may be a water bath in which crates of the food are heated to a pre-set temperature, and then cooled by draining and adding cold water. A continuous version may convey containers through a hot water batch followed by a cold water bath. Steam tunnels may also be used with the advantage of faster heating, resulting in shorter residence time and smaller equipment. Temperatures in the heating zones may be controlled depending on the amount of air present. Acid products such as fruit or acidified vegetables like beetroot can be pasteurized in a retort.

Sterilisation

Unlike pasteurized products where the survival of heat resistant microorganisms is accepted, the aim of sterilization is the destruction of all bacteria including their spores. Heat treatment of such products must be severe enough to inactivate/kill the most heat resistant bacterial microorganisms, which are the spores of Bacillus and Clostridium. Food products filled in sealed containers are exposed to temperatures above 100 °C in pressure cookers. Temperatures above 100 °C, usually ranging from 110-121 °C depending on the type of product, must be reached inside the product. Products are kept for a defined period of time at temperature levels required for the sterilization depending on type of product and size of container.

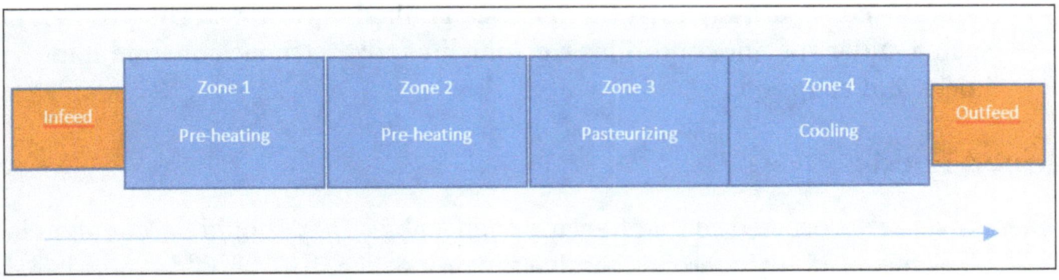

If spores are not completely inactivated, vegetative microorganisms will grow from the spores as soon as conditions are favourable again. Favourable conditions will exist when the heat treatment is completed and the products are stored under ambient temperatures. The surviving microorganisms can either spoil preserved food or produce toxins which cause food poisoning. Amongst the two groups of spore producing microorganisms Clostridium is more heat resistant than Bacillus. Temperatures of 110 °C will kill most Bacillus spores within a short time. In the case of Clostridium temperatures of up to 121 °C are needed to kill the spores within a relatively short time. These sterilization temperatures are needed for short-term inactivation (within a few seconds) of spores of Bacillus or Clostridium. These spores can also be killed at slightly lower temperatures, but longer heat treatment periods must be applied.

From the microbial point of view, it would be ideal to employ very intensive heat treatment which would eliminate the risk of any surviving microorganisms. However, most food products cannot be submitted to such intensive heat stress without suffering degradation of their sensory quality or loss of nutritional value (destruction of vitamins and protein components). In order to comply with above aspects, a compromise has to be reached in order to keep the heat sterilization intensive enough for the microbiological safety of the products and as moderate as possible for product quality reasons.

"Commercial sterility" implies less than absolute destruction of all micro-organisms and spores, but any remaining would be incapable of growth in the food under existing conditions. Time-temperature combination required to inactivate most heat resistant pathogens and spoilage organisms. Most heat resistant pathogen is Clostridium botulinum. Most heat-resistant (non-pathogenic) spoilage microorganisms are Bacillus stearothermophilis and Clostridium thermosaccharolytom. Severity of treatment can result in substantial changes to nutritive and sensory characteristics. Two typical forms of sterilised product are:

- In package sterilised, in which product is packed into containers and the container of product is then sterilised e.g. canning, some bottled products, retort pouches.
- UHT or Aseptically processed products in which the product and the package is sterilised separately then the package is filled with the sterile product and sealed under specific conditions e.g. long life milk, tetrapack or combibloc fruit juices and soups etc.

Canned Foods

Canned foods are processed so that they are shelf stable. They should be 'commercially sterile'. That means if any microbes survive the processing, they should not be capable of growing (and therefore spoiling the contents) under the normal storage conditions of the can. Most canned foods are sterile (i.e. there are no living organisms present) but some may contain viable organisms which cannot grow because of unsuitable conditions e.g.

- Water
- Temperature
- pH

- Water activity
- Preservatives

If a canned food is spoilt by microbial spoilage, examination of the microbial types that caused it can pinpoint the offending errors in processing or handling.

Conditions Affecting the Growth of Microorganisms

Water

Water content and the availability of water Aw can affect the growth of microbes in food.

Temperature

Temperature influences the rate of growth of microbes as well as determining which microbes will grow. Microbes grow fastest at their optimum temperature. For convenience microbes can be divided into groups which have similar optimum temperature for growth.

Table: Growth temperatures (°C) for microbial growth:

Group	Min.	Opt.	Max.
Thermophiles	40	55	75
Mesophiles	5	37	45
Psychotrophs	-3	20	30

Oxygen Requirements

Micro-organisms can be classified into three general groups regarding their oxygen requirements.

- Aerobes – Can only grow in the presence of oxygen.
- Anaerobes – Can only grow in the absence of oxygen.
- Facultative Anaerobes – Adaptable. Grows best aerobically but can grow anaerobically.

pH

In regard to pH, microbes have ideal pH ranges within which they grow as follows:

Table: pH ranges for microbial growth.

Group	pH
Low acid	>5.0
Medium acid	4.5 - 5.0
Acid	3.7 - 4.5
High acid	< 3.7

Types of Microorganisms Important in Retorted Foods

A number of organisms are important when it comes to the safe processing of canned foods.

Table: Microorganisms important for retorted foods.

Type	Species	Description
Thermophilic Spore Formers.	Flat Sours - B. sterothermophilus.	High heat resistance, product acid, don't produce gas, found in sugar, salt and spices.
	Thermophilic Anaerobes – C.thermosaccharolyticum.	High heat resistance, product acid and gas (CO_2).
	Sulphide types – Desutfomotomaculum nigrificans.	High heat resistance, produce H_2S.
Mesophilic Spore Formers.	C.sporogenes, C.botulinium.	Produce gas (CO_2 and sometimes H_2, moderate heat resistance.
(The process should be designed to kill these microbes).	Bacillus spp – B.polymyxa, B.macerans etc.	Moderate to low heat resistance, some may grow in acid foods.
Non Spore Forming Microbes.	Various.	Occur only in grossly under processed or leaking caps. Can be almost any microbe depending on acidity of the product. May or may not produce gas Usually in mixed populations.

Microbial Spoilage of Canned Foods

There are a number of important factors which can cause spoilage of canned foods. Table: Factors affecting spoilage of canned foods.

Type	Description
Pre-process spoilage.	Delays between filling and retorting can allow microbes to grow and produce gas or spoil food. Retorting kills microbes but the can will be swollen and food spoilage.
Not processed.	Filled cans missing retort.

Under processed.	Caused by: • Incorrect calculations. • Faulty retort operation. • Operator error e.g. inadequate venting. • Poor retort design e.g. cold spots. • Higher spore load – poor or different raw ingredients. Under processing usually still kills vegetative cells. Survivors are usually mesophilic spore formers or moderate heat resistance.
Thermophilic Spoilage.	Canning operations are sometimes not designed to kill thermophiles of high heat resistance (e.g. B. stearothermophilus of D 121.1 = 5 min) as they do not grow below 40 °C. If they survive they will grow if there is either slow cooling or storage at high temperatures. Thermophilic spore formers will be found in pure cultures.
Leaker Spoilage.	If can seams are inadequately formed, microbes may enter can after processing, particularly when the can is moist e.g. during cooling. Usual contamination is a mixed of a variety of non-heat resistant microbes. Cans may leak food or if leakage point is blocked with food, they can swell.

Sterilisation Process and Equipment

The sterilization process in the canned product can be subdivided into three phases. By means of a heating medium (water or steam) the product temperature is increased from ambient to the required sterilization temperature (phase 1 = heating phase). This temperature is maintained for a defined time (phase 2 = holding phase). In (phase 3 = cooling phase) the temperature in the can is decreased by introduction of cold water into the autoclave.

Autoclaves or Retorts

In order to reach temperatures above 100 °C ("sterilization"), the thermal treatment has to be performed under pressure in pressure cookers, also called autoclaves or retorts.

In autoclaves or retorts, high temperatures are generated either by direct steam injection, by heating water up to temperatures over 100 °C or by combined steam and water heating. The autoclave must be fitted with a thermometer, pressure gauge, pressure relief valve, vent to manually release pressure, safety relief valve where steam is released when reaching a certain pressure, water supply valve and a steam supply valve. The steam supply valve is applicable when the autoclave is run with steam as the sterilization medium or when steam is used for heating up the sterilization medium water.

Simple Small Autoclaves

These are usually vertical autoclaves with the lid on top. Through the opened lid the goods to be sterilized are loaded into the autoclave. The cans are normally placed in metal baskets. The baskets are placed in the autoclave, either singly or several stapled on top of each other. Before starting the sterilization, the lid must be firmly locked onto the body of the autoclave. The autoclave and lid are designed to withstand pressures up to 5.0 bar. These types of autoclaves are best suited for smaller operations as they do not require complicated supply lines and should be available at affordable prices.

Larger Autoclaves

These are usually horizontal and loaded through a front lid. Horizontal autoclaves can be built as single or double vessel system. The double vessel systems have the advantage that the water is heated up in the upper vessel to the sterilization temperature and released into the lower (processing) vessel, when it is loaded and hermetically closed. Using the two–vessel system, the heat treatment can begin immediately without lengthy heating up of the processing vessel and the hot water can be recycled afterwards for immediate use in the following sterilization cycle.

If steam is used instead of water as the sterilization medium, the injection of steam into a single vessel autoclave will instantly build up the autoclave temperature desired for the process.

Rotary Autoclaves

Another technology employed is rotary autoclaves in which the basket containing the cans rotates during sterilization. This technique is useful for cans with liquid or semi-liquid content as it achieves a mixing effect of the liquid/semi-liquid goods resulting in accelerated heat penetration. The sterilization process can be kept shorter and better sensory quality of the goods is ensured.

At the final stage of the sterilization process the products must be cooled down as quickly as possible. This operation is done in the autoclave by introducing cold water. The contact of cold water with steam causes the latter to condense with a rapid pressure drop in the retort. However, the overpressure built up during thermal treatment within the cans, jars or pouches remains for a certain period.

During this phase, when the outside pressure is low but the pressure inside the containers is still high due to high temperatures there, the pressure difference may induce permanent deformation of the containers.

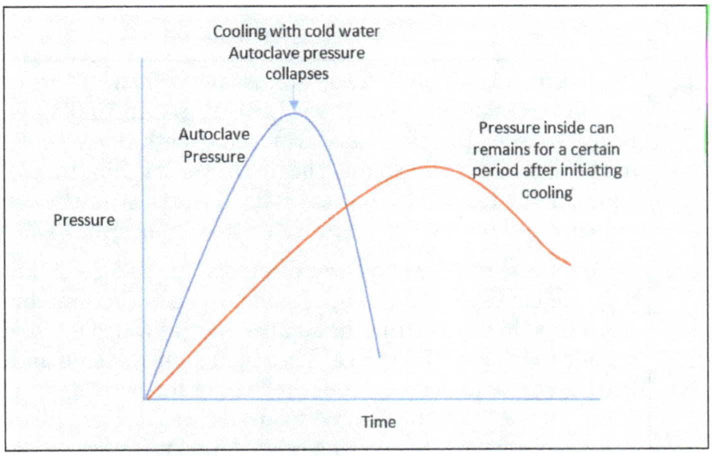

Pressure inside autoclave (blue) and inside cans (red) during heating and cooling phase.

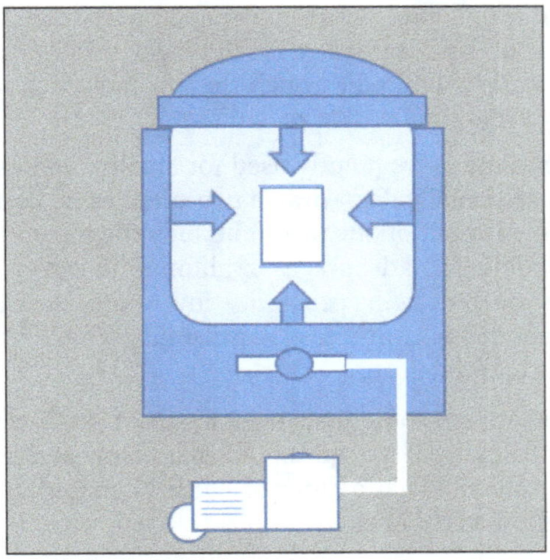

Producing counter pressure on cans inside the autoclave with compressed air.

Therefore, high pressure difference between the autoclave and the thermal pressure in the containers must be avoided. This is generally achieved by a blast of compressed air into the autoclave at the initial phase of the coolings Suffcient hydrostatic pressure of the introduced cooling water can also build up counter pressure so that in specific cases, in particular where strong resistant metallic cans are used, the water pressure can be suffcie and compressed air may not be needed. For the stabilization of metallic cans, stabilization rims can be moulded in lids, bottom and bodies.

Types of Containers for Thermally Treated Products

Containers for heat-preserved food must be hermetically sealed and airtight to avoid recontamination from environmental microflora. Most of the thermally preserved products are in metal containers (cans). Others are packed in glass jars or plastic or aluminium/ plastic laminated pouches.

Type	Description
Metal containers are cans or "tins"	Produced from tinplate. They are usually cylindrical. However, other shapes such as rectangular or pear-shaped cans also exist. Tinplate consists of steel plate which is electrolytically coated with tin on both sides. The steel body is usually 0.22 to 0.28mm in thickness. The tin elaryer is v thin (from 0.38 to 3.08 µm). In addition, the interior of the cans is lined with a synthetic compound to prevent any chemical reaction of the tinplate with the enclosed food. Tin cans consist of two or three elements. In the case of three-piece steel cans, they are composed of the body and two ends (bottom and lid). The body is made of a thin steel strip, the smaller ends of which are soldered together to a cylindrical shape. Modern cans are induction-soldered and the soldering area is covered inside with aside-strip coating for protection and coverage of the seam. The use of lead soldered food cans was stopped decades ago. Hence the risk of poisonous lead entering canned food no longer exists. Two-piece steel cans have a lid similar to the three-piece cans but the bottom and body consist of one piece, which is moulded from a circular flat piece of metal into a cup. These cup-shaped parts may be shallow-drawn (with short side wall) or deep-drawn (with longer side walls). However, the length of the side walls is limited through the low moulding ability of steel (example: tuna tins 42/85mm, i.e. side wall: diameter =1:2). Aluminium is frequently used for smaller and easy-to-open cans. Aluminium cans are usually deep-drawn two-piece cans, i.e. the body and the bottom end are formed out of one piece and only the top end is seamed on after the filling operation. The advantages of aluminium cans compared to tin cans are their better deep-drawing capability, low weight, resistance to corrosion, good thermal conductivity and easy recyclability. They are less rigid but more expensive than steel plate cans.
Glass jars	Glass jars are sometimes used for meat products but are not common due to their fragility. They consist of a glass body and a metal lid. The seaming panel of the metal lid has a lining of synthetic material. Glass lids on jars are fitted by means of a rubber ring.
Retortable pouches	Retortable pouches, which are containers made either of laminates of synthetic materials only or laminates of aluminium foil with synthetic materials, are of growing importance in thermal food preservation. Thermo-stabilized laminated food pouches, have a seal layer which is usually PP (polypropylene) or PP-PE (polyethylene) polymer, and the outside layers are usually made of polyester (PETP) or nylon. They can be used for frankfurters in brine, ready-to-eat meat dishes etc. From certain laminated films, for instance, polyester / polyethylene (PETP/ PE) or polyamide/polyethylene (PA/ PE), relatively rigid container can be made, usually by deep drawing. They are used for pieces of cured ham or other kinds of processed meat. Small can-shaped round containers are made from aluminium foil and polyethylene (PE) or polypropylene (PP) laminate and are widely used for small portions, particularly of sausage mix. PE or PP permits the heat-sealing of the lid made of the same laminate onto these containers, which can then be subjected to intensive heat treatment of 125 °C or above. One advantage of the retortable pouches/laminated containers is their good thermal conductivity which can considerably reduce the required heat treatment time and hence is beneficial for the sensory product quality.

Cleaning of Containers Prior to Filling

Rigid containers (cans, glass jars) are delivered open to meat processing plants, i.e. with the lids separate. During transport and storage, dust can settle inside the cans, which must be removed prior to filling the cans. This can be done at the small-scale level by manually washing the cans with hot water. Industrial production canning lines are equipped with steam cleaning facilities, where steam is blown into the cans prior to filling.

Seaming of Cans

After the can is filled with the product mix the can is sealed with a tight mechanical structure - the so-called double seam. The double seam, in its final form and shape, consists of three layers of lid (D, black colour) and two layers of body material (D, striated). The layers must overlap significantly and all curves must be of rounded shape to avoid small cracks. Each double seam is achieved in two unit operations referred to as "first operation" (A, B) and "second operation" (C, D).

The can covered with the lid is placed on the base plate of the can seaming machine. The can is moved upwards while the seaming chuck keeps the lid fixed in position. The pressure applied to the can from the base plate can be regulated and must be strong enough to ensure simultaneous movement of the lid and the can to avoid scratching-off of the sealing compound.

In the first operation the lid hook and body hook are interlocked by rolling the two into each other using the seaming roll with the deep and narrow groove. The body hook is now almost parallel to the lid hook and the curl of the lid adjacent to or touching the body wall of the can. In the second operation, the interlocked hooks are pressed together by a seaming roll with a flat and wide groove. Wrinkles are ironed out and the rubber-based material is equally distributed in the seam, filling all existing gaps thus resulting in a hermetically sealed container.

Design of Seaming Rolls

The seaming rolls for the first and second operations are designed differently in order to facilitate the respective operations. The seaming roll for the first operation has a deep but narrow groove to interlock body and lid hock (rolling the hocks into each other). The seaming roll for the second operation has a flat but wide groove to press the interlocked hooks together (sealing the seam). The first action first roll) is rolling (interlocking) the hooks, the second action (second roll) is compressing (sealing) the seam.

Death Rate Curve (D value)

At slightly elevated temperatures most microbes will grow and multiply quickly. At relatively high temperatures, microbes can be destroyed. However, there is a lot of variation within any one population of microbes of the same species – most will be killed relatively quickly, others can survive much longer. If a population of microbes is held at a constant high temperature, the number of surviving spores or cells plotted against time (on a logarithmic scale) will look like the following graph – which is referred to as the 'death rate curve'.

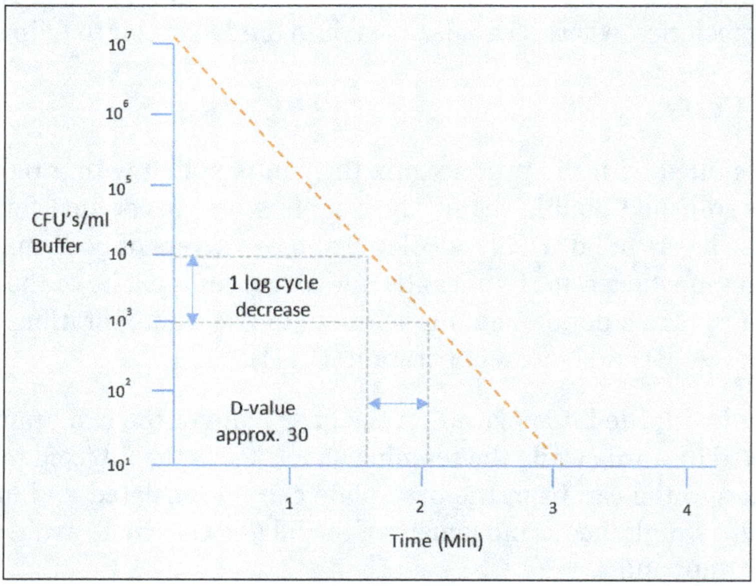

Death rate curve (D-value).

This graph is a straight line – it is referred to as the Logarithmic order of death. Logarithms refers to the power to which a base must be raised to produce a given number. For example, if the base is 10, then the logarithm of 1,000 (written log 1,000 or log10 1,000) is 3 because 103 = 1,000. The "death rate curve" is a straight line when plotted using a logarithmic scale – this means that if in some time period the number was reduced from 1000 to 100 (divided by ten, sometimes referred to as "1 log reduction"), then if you had held the microbes at the same temperature for twice that time period, the number would have been reduced to 1 (divided by 100, or "2 log reductions").

The time period for each "log reduction" is referred to as the decimal reduction time or D value. For example the D-value of Bacillus stearothermophillus a common spoilage microorganism at 121 °C is about 4 minutes. This means if you had cans of food product each containing 1000 of these spore and you held the product at a constant temperature of 121 °C.

- After 4 minutes (1 D-value) there would be 100 spores surviving in each can (1 log reduction).
- After 8 minutes (twice D-value) there would be 10 spores surviving in each can (2 log reductions).
- After 12 minutes (3 times D-value) there would be 1 spore surviving in each can (3 log reductions).

If this food product, with an initial count 1000 spores of Bacillus stearothermophillus, was held for 16 minutes at 121 °C it would result in 4 log reductions, or 0.1 spores surviving in each can. 0.1 spores per can means that on average there would be one spore surviving in each group of ten cans. After holding for 20 minutes there would be one spore per 100 cans and so on.

Based on this:

- The higher the number of microbes initially present the longer it takes to reduce the numbers to an acceptable level. Therefore, good quality raw materials and hygienic pre-processing is essential if the commercial sterility of the processed product is to be assured.
- It is theoretically impossible to destroy all cells – therefore we reduce the probability of spoilage to an acceptable small number – perhaps 1 in 1 million. The probability of a pathogen surviving must be even less – perhaps on in one billion or less.
- The above refers to holding the product at a constant temperature. Remember destruction of microbes is temperature dependent – they get killed more quickly at the higher temperatures. Therefore you would expect that if you increase the temperature, decimal reduction time D-value) would decrease.

Thermal Death Time (TDT) Curve

If D-value versus time is plotted – again on a logarithmic scale, the graph looks very similar to the one previously. This one is called the Thermal death time (TDT) curve. This time the straight line graph means that if you change the temperature by a certain amount, the D-value will change by a factor of 10. If you had changed it by twice that amount, D-value will change by a factor of 100. The change in temperature to cause a factor of the ten change in D-value is referred to as that z-value.

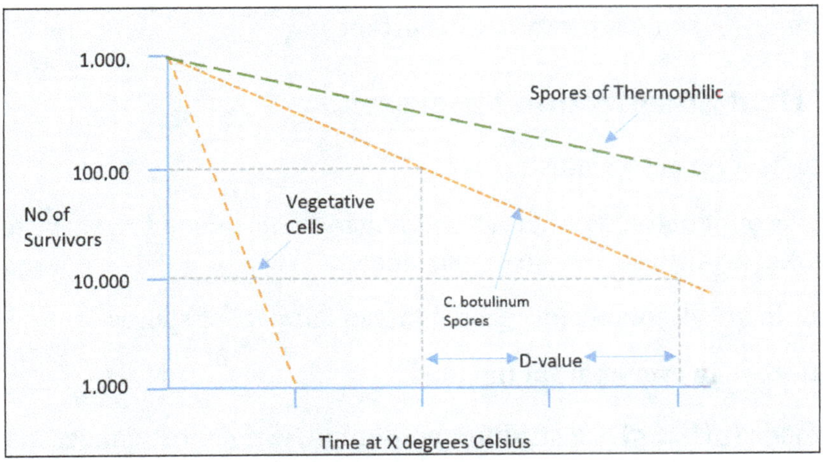

Thermal death rate curves.

The z-value for Bacillus stearothermophillus is 10 °C. Remember the D-value for this microorganism at 121 °C is 4 minutes. Therefore if you held the containing this microbe at 111 oC (10 oC, or one z-value, less than 121 oC), D-value would be 400 minutes.

That is, for Bacillus stearothermophillus, 4 minutes at 121 °C will have the same effect (one log reduction in spores) as 40 minutes at 111 °C, which would have the same effect as 400 minutes at 101 °C. It is obvious why using high processing temperatures is an advantage. The D-values of different microbes differ greatly – for example, the D-value of Clostridium botulinum at 121 °C is about 0.21 minutes. However the z-value of microorganisms is close to 10 °C.

Some Factors that Affect the Heat Resistance of Micro-organisms

A range of factors affect the heat resistance of micro-organisms. The most important are:

- Type of Micro-organism – species and strains differ, spores are more resistant than vegetative cells.
- Formation Conditions during cell growth or spore – e.g. spores produced at higher temperature are more heat resistant, stage of growth and the type of medium in which they grow can also affect heat resistance.
- Conditions during Heat Treatment including pH. Pathogenic and spoilage bacteria are less heat resistant at more acid (low) pH, yeasts and fungi are more acid tolerant but less heat resistant than bacterial spores.
- A_w – moist heat is more effective than dry heat.
- Composition – e.g. protein, fats and high concentration of sucrose increases heat resistance.
- D and z-values of enzymes are generally in a similar range to those of micro-organisms, but some are very heat resistant.

Design of Heat Sterilization Processes

The design of heat processes must:

- Take account of the type of microorganism (determined largely by food conductions e.g. acidity) and its heat resistance.
- Result in an acceptably low probability of survival of spores.
- Be effective in every part of the food.

In low acid foods (pH<4.5), Clostridium botulinum, is the most dangerous, heat resistant spore forming pathogen (D121=0.1 to 0.2 min). It is anaerobic and so can survive and grow in a sealed can. Its destruction is a minimum requirement of heat sterilisation. This is often interpreted as "12D" process – that is, the product must be treated for 12 times the D-value of the microbe. For Clostridium botulinum this is a process equivalent to about 2.5 minutes at 121 °C – this is commonly known as a "botulinum cook". Normally a more severe heat treatment is required to destroy other more heat resistant spoilage bacterial. For example Bacillus stearothermophillus (thermophile – won't grow at less than 35 °C, so proper can cooling is important) can produce the "flat sour" defect. Its D-value at 121 °C is commonly around 4 min, but it is not of health significance.

In high acid foods (pH<4.5), the anaerobic pathogens cannot grow or produce toxins. Spoilage microorganisms are quickly killed at temperatures of about 90 °C. Therefore the minimum treatment applied to high acid foods often involves ensuring every part of the product reaches a temperature of at least 95 °C e.g. pasteurisation. In acid foods where the pH is close to 4.5 (e.g. foods such as tomatoes and pears) Clostridium butyricum can cause spoilage. It is a common soil borne micro-organism, and grows easily on surfaces in the food plant. It is not killed by processes commonly used for acid foods and can cause swelling/bursting of the cans in about 2 weeks.

F_o Value

The amount of heat treatment applied to a food product can be measured using the F-value-concept. This concept is practiced in canning plants, in particular as part of the HACCP-system. The size and format of cans is of utmost importance for the speed of heat penetration. Temperatures to be achieved at the "cold point" of the can where the heat arrives last, are reached faster in small cans due to the shorter distance to the heat source than in large cans.

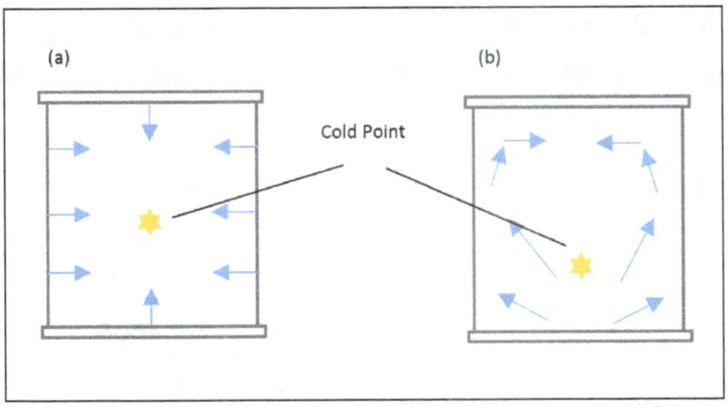

The F_0 value is a measure of the "sterilising value" of a process. It can be thought of as the time required at a temperature of 121 °C to reduce microbial numbers by the same amount as the actual process being considered.

Remember processes are not always carried out at 121 °C and certainly product temperature is not constant at this temperature throughout the process.

It therefore provides a basis for comparing different heat sterilisation procedures if two processes have the same The F_0 value, they provide the same level of sterilisation.

The temperature of 121 °C is simply an arbitrary reference – there is nothing special about this particular temperature. Why choose and off temperature like 121 °C? In the past someone decided 250°F which is equal to 121 °C was a good reference temperature. More accurately it is 121.1 °C.

A similar concept to F_0 often used in determining the heat treatment of beers and other high acid foods is "pasteurising units" (or PU's) – 1 PU is equivalent to pasteurising at 60 °C for one minute. The minimum treatment for low acid products, the "botulinum cook", therefore has a F_0 of 2.5 minutes (i.e. 12* 0.21 = 2.5 min).

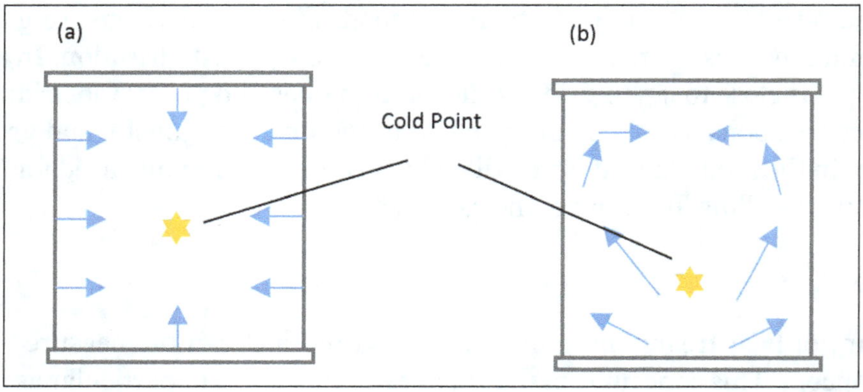

The required level of heat treatment (F_0 of the process) may vary with factors such as pH and carbohydrate level, and type and expected level of contamination with micro-organisms. Other chemical additives may also assist inhibition of microorganisms e.g. salt, alcohol, nitrite and misin (these last two are both 'sporostatic' and stop spores germinating and so enable the use of lesser processing conditions). Also some products require additional processing to achieve the required level of cook e.g. baked beans must be soft enough.

Table: F-values (per minute) for the temperature range of 100 °C to 135 °C.

°C	F-value	°C	F-value
100	0.0077	118	0.4885
101	0.0097	119	0.6150
102	0.0123	120	0.7746
103	0.0154	121	1
104	0.0194	122	1.2270
105	0.0245	123	1.5446
106	0.0308	124	1.9444

107	0.0489	125	2.4480
108	0.0615	126	3.0817
109	0.0775	127	3.8805
110	0.0975	128	4.8852
111	0.1227	129	6.1501
112	0.1545	130	7.459
113	0.1545	131	9.7466
114	0.1945	132	12.2699
115	0.2449	133	15.4560

Lethality Factor "L"

Given that the F_0 is based on a constant reference temperature of 121 °C, but the product is mostly at a different temperature, how can the F_0 be calculated? This is the purpose of the Lethality Factor or "L-value". It is defined as the time at 121.1 °C which is equivalent in sterilising value to one minute at some other temperature. One minute at some temperature will contribute "L" minutes worth of F_0, where "L" is the Lo value for the temperature concerned. The L-value is dependent on the z-value of the micro-organism being considered, but for most purposes z=10 °C. L-value can be calculated from the formula or can be read from a table.

$$L = 10(T-121.1)/z.$$

An example – A product is held at a temperature of 118 °C for a period of 12 minutes. Ignoring other heating and cooling periods, what is the F_0 value of this process? From the formula, the L-value for 118 °C is 0.490. That is each minute at 118 °C contribures 0.490 minutes to the F_0 value. Therefore the F_0 value of this process = 12 x 0.490 = 5.9 minutes. Calculating the F_0 value when temperatures vary.

In a real retort process the temperature of the product is not constant – it slowly heats up, will stay at a constant temperature for some time, then cool down again. The period when the product is heating and cooling contribute significantly to the severity of the process. To calculate the F_0 value of such a process, the contribution of the varying temperatures must be converted to an equivalent F_0 value. This is achieved based on the L-value, as indicated previously.

Graphical Method

This involves drawing a graph of the product temperature vs time, then looking up the L-value of each temperature, and plotting L-value against time. The area under this graph is a measure of the L-value.

Trapezoidal Integration or General Method

For this method, determine the L-value for each temperature measurement, add the L-value together then multiply by the time interval in minutes between temperature measurements (if temperatures are measured every minute there is no need to multiply). Obviously as the severity of the process is related to the time spent at high temperatures the faster a product is heated the greater will be the severity of the process (for the same process time).

A number of factors affect the rate at which a product heats inside a container:

- Type of container – For example glass is not a good conductor of heat so you would expect product in a glass jar to heat more slowly than an equivalent size/ shape metal can.
- Size and shape of the container – Obviously a large container will take longer to heat than a small container.
- Retort temperature – A higher retort temperate will result in more rapid heating but also may lead to more over processing of product near the package surface.
- Agitation of the containers will increase the heating rate by mixing the contents of the container, especially with viscous or semi-solid foods. End over end agitation is better than axial agitation).
- Type of product – Obviously different products conduct heat more or less easily and have different heat capacities. Some products are more viscous than others which can have a particularly significant effect in agitating retorts. Therefore different products will heat at a different rate.
- Headspace – Insufficient headspace can also affect the rate of heating, especially in an agitating retort.

Therefore if any of these factors change, the severity of the process needs to be reevaluated.

Non-thermal Techniques

The food-processing industry has made large investments in processing facilities relying mostly on conventional thermal processing technologies with well-established reliability and efficacy Food contains many heat sensitive nutrients which include vitamins, minerals, and nutrients having functional properties such as pigments, antioxidant, Bioactive compounds. Many processes during manufacturing of food cause detrimental effects on these nutrients. Retention of these nutrients in food products requires innovative approaches for process design because of their sensitivity to a variety of physical and chemical factors, which causes either loss of biological functionality, chemical degradation and premature or incomplete release. Alternative methods for thermal processing of food are gaining importance, due to increased consumer demand for new methods of food processing that have a reduced impact on nutritional content and overall food quality. Also as a result of the increasing consumer demand for minimally-processed fresh-like food products with high sensory and nutritional qualities, there is a growing interest in non-thermal processes for food processing and preservation.

Non-thermal food processing/preservation methods interest food and food packaging scientists, manufacturers and consumers because they exert a minimal impact on the nutritional and sensory properties of foods, and extend shelf life by inhibiting or killing microorganisms. They are also considered to be more energy efficient and to preserve better quality attributes than conventional thermally based processes. Non-thermal physical processes are evolving as potential alternatives to thermal and chemical unit operations in food processing. Non-thermal methods allow the processing of foods below temperatures used during thermal pasteurisation, so flavours, essential nutrients, and vitamins undergo minimal or no changes. These processes also meet industry needs by offering value- added products, new market opportunities and added safety margins.

Non-thermal inactivation techniques have been a major research issue, driven by an increased consumer demand for nutritious, fresh like food products with a high organoleptical quality and an acceptable shelf life. Investigated inactivation technologies are ionisation radiation, HHP, pulsed electrical fields, high pressure homogenisation, UV decontamination, pulsed high intensity light, high intensity laser, pulsed white light, high power ultrasound, oscillating magnetic fields, high voltage arc discharge and streamer plasma.

Pulsed X-ray

Electrons have a limited penetration depth of about 5 cm in food, while X-rays have significantly higher penetration depths (60 - 400 cm) depending upon the energy used. Pulsed X-rays are generated using radionuclide sources that utilizes a solid state-opening switch to generate electron beam X-ray pulses of high intensity. The radionuclides Co-60 and Cs-137 are produced by neutron bombardment of Co-59 and Cs-136 as a fission fragment of a nuclear power reactor operation. They emit γ-radiation of discrete energy. These radionuclide sources require permanent massive concrete shielding to protect workers and the environment from their permanent radiation. Second approach is electrically driven radiation sources that switch off when the radiation is no longer needed are easier to incorporate into existing food processing plant.

Linear Induction Electron Acceleration (LIEA) generates broad spectrum ionizing radiation by targeting the accelerated electron beam to collide with a heavy metal converter plate. This plate converts the electron beam in X-rays with a broadband photon-energy spectrum. Then, by filtering the energy spectrum of the radiation, high-energy, highly penetrating radiation is produced, resulting in smaller variations in dose uniformity of food packages and higher quality. LIEA can deliver dose rates many orders of magnitude higher than possible with Co-60 sources. Consequently, ultra-short, highintensity radiation treatments can be applied, resulting in higher local radical concentrations and favouring radicalradical recombination reactions. This reduces the diffusion of radical species, which are thought to be responsible for undesirable effects of irradiation on food quality.

Salient Features of this technology are (1) flexibility of controlling the direction of the electrically produced radiation; (2) the flexibility of shaping the geometry of the radiation field to accommodate different package sizes; and (3) its high reproducibility and versatility. The kinetic energy limit for Xray irradiation is 5MeV (Codex Alimentarius General Standard on Food Irradiation. X-ray treatment reduces or eliminates Salmonella serovars in poultry, mold growth on strawberries and sprout development in potatoes. Salmonella serovars have been found to be the most radiation sensitive of all pathogenic organisms on foods. As a method of food preservation, X-ray treatment has low energy requirements. Microbial inactivation by all types of ionizing radiation is thought to happen through 2 main mechanisms:

Direct interaction of the radiation with cell components and indirect action from radiolytic products, such as the water radicals. The primary target of ionizing radiation appears to be chromosomal DNA, although effects on the cytoplasmic membrane may also play a role. Changes to chromosomal DNA and cytoplasmic membrane can cause microbial inactivation or growth inhibition. Many studies have shown that ions, excited atoms and molecules generated during irradiation have no toxic effect on humans USA - FDA Center for Food Safety and Applied Nutrition.

Pulsed Visible Light

The technique of pulsed light food processing was developed as a non-thermal food processing technique, that involves discharge of high voltage electric pulses (upto 70 Kilovolt/cm) into the food product placed between two electrodes for few seconds. It is one of the emerging technologies which are used for the replacement of traditional thermal pasteurization among non thermal processes. It is a decontamination technique which aims at reducing the pests, spoilage microorganisms and pathogens from food without much effect on its quality. It is recognized by several names in scientific literature i.e., Pulsed ultraviolet light, high intensity broad-spectrum pulsed light, Pulsed light and pulsed white light. The pulsed light processing can be described as a sterilization or decontamination technique used mainly to inactivate surface micro-organisms on foods, packaging material and equipments. This technique uses light energy in concentrated form and exposes the substrate to intense short bursts of light (pulses). Typically for food processing about one to twenty flashes per second are applied.

Mechanism of Microbial Inactivation

The lethality of Pulsed Light may be attributed to its rich broad spectrum ultraviolet content, its short duration, high peak power and the ability to regulate the pulse duration and frequency output of flash lamps. As a substantial portion of the Pulsed light spectrum covers ultraviolet light, it is considered that ultraviolet plays a vital role in the microbial cell inactivation. It was also found that that there is no killing effect if a filter is used to remove ultraviolet (UV) wavelength region lower than 320 nm. The ultraviolet spectrum comprises of three wave ranges: Long-wave ultraviolet -A (320-400 nm), Medium- wave ultraviolet -B (280-320 nm) and Short-wave ultraviolet -C (200-280 nm). Mechanisms that have been proposed to explain the lethality of pulsed light treatment are related to ultraviolet (UV) part of the spectrum which include photochemical and photothermal effect. The lethal effect of pulsed light can be due to photochemical or photothermal mechanism or both may exist simultaneously.

However, their relative importance depends on the fluence and target microorganism. The lethal effect of pulsed light was explained by most of the authors on the basis of photochemical mechanism e.g., the inactivation achieved by was associated with less than 10 °C rise in temperature concluded that the lethality can be attributed to the photochemical action of the shorter ultraviolet wavelengths. The primary target cell of pulsed light in photochemical mechanism is nucleic acid as DNA is the target cell for these ultraviolet wavelengths. Ultraviolet light absorbed by the conjugated carbon-carbon double bonds in proteins and nucleic acids induces the antimicrobial effect as it changes the DNA and RNA structures. The bactericidal effect is attributed to the high energy short wave ultraviolet-C range. In the ultraviolet-C range of 250-260 nm, alterations in DNA take place due to pyrimidine dimers mainly thymine dimers. These dimers inhibit the formation of new DNA chains in the process of cell replication resulting in the chologenic death of affected microorganisms by ultraviolet.

The ultraviolet-C treatment of bacterial spores may result in the formation of spore photo-product 5-thyminyl-5, 6-dihydrothymine and in single-strand breaks, double-strand breaks and cyclobutane pyrimidine dimers. It was also found by experiments that enzymatic repair of DNA does not occur after damaged by pulsed light. The lethal effect of Pulsed light can also be due to photothermal effect. Proposed that with a fluence exceeding 0.5 Joule/cm2, the disinfection is achieved through a rupture of bacteria during their temporary overheating caused by absorption of all ultraviolet light from a flash lamp. This hypothesis become evident by when they showed electron-microscope photographs of flashed Aspergillusniger spores presenting severe deformation and rupture. The ruptured top of spore become evident of an escape of an overheated content of the spore, which became empty after such an internal "explosion" and "evacuation" of its content took place during the light pulse. As Pulsed light causes cell membrane damage, it could be considered as a technique for sterilization, reported that PL treatments achieves high levels of microbial inactivation on relatively simple surfaces, while generally showed only 1–3 log reductions on complex surfaces such as meats. Part of the radiation may have been absorbed by proteins and lipids, thus decreasing the effective radiation dose on microorganisms. Proteins have strong absorption of UV at about 280 nm as well as at higher wavelengths of the UV-B region, while lipids with isolated or conjugated double bonds also absorb UV demonstrated that beef steaks treated with PL using 5 J cm^{-2} to each side and stored 3 days at 4–5 °C exhibited 2 log reductions in microbial counts. Listeria innocua was reported to be reduced by 2 log cycles on hot dogs after a PL treatment. Milk was efficiently cold pasteurized by the exposure to PL at a minimum dose of 12.6 J cm^{-2} delivered in 56 s. Complete inactivation of S. aureus was obtained when processing milk in a continuous system applying PL.

Advantages: The intensity of light, that lasts for only a second, is 20,000 times brighter than sunlight, but there is no thermal effect, so quality and nutrient content are retained The xenonflash lamps used in pulsed light treatment are more ecofriendly than the mercury vapour lamps used in ultraviolet (UV) treatment.

Disadvantages: A possible problem of this preservation method is that folds or fissures in the food may protect microbes from being exposed to the pulsed light. There might be some strains of micro-organisms which are resistant to the pulsed light treatment, for example Listeria monocytogenes. This technique for decontamination of micro-organisms is useful mostly in case of liquid foods and surface of solid foods and hence limiting its application.

Pulsed Electric Field (PEF)

During PEF processing, energy is stored in a capacitor, retrieved from a high-voltage power supply, and is discharged through foods that are either static or are flowing through a treatment chamber. PEF uses short bursts of electricity (submicroseconds to milliseconds), yielding few to no detrimental affects on quality attributes in pumpable foods. This process pulses high voltage (10–80 kV/cm) into foods placed between two electrodes, for less than one second, near ambient temperature, then packaged aseptically and distributed refrigerated. This process attains a 5 log reduction on most pathogenic bacteria by rupturing the cell membranes in liquid media. It causes only minimal detrimental changes to the physical and sensory properties in foods, helps retain 'fresh' quality and assists in nutrient retention. PEF can be applied to the pasteurization of liquid products, in continuous systems, such as milk, yogurt, juices, liquid eggs, soups, brines and other products that can withstand high electric fields. High electric field pulses can be employed to aid in the extraction of polysaccharides and peptides. PEF has limited effects on microbial spores, cannot be used on products that contain or could form air bubbles, and cannot be used on foods that have higher or variable electrical conductivity. 'Pressure is applied to inhibit the formation of air bubbles in which electrical arcing could occur with fields above 20000V/cm'. Since PEF kills cells and impairs water retention, it can aid in filtration methods and can also be used for the extraction of sugars and starches from root vegetables. PEF only affects a few enzymes, a concern in the juice industry. Enzymes negatively affect juice processing by reducing pectin, which aids in fruit particle suspension, and may cause sedimentation, discolouration and flavour degradation. Critical factors that can affect the inactivation of microorganisms using PEF include process variables, media and microbial factors. PEF processing variables include pulse wave and width, electric field intensity, temperature and time of exposure. Electric fields are produced on equipment that can be compared to that of radar. The most typical equipment generates a short square wave and reverses polarity, in part to avoid erosion of electrodes. Two other wave forms can be produced, which include sinusoidal and exponential decay. In reverse polarity, bipolar generators are twice the cost of non-polar units, making this an expensive process at this time. One operation has received FDA approval and is being used by Genesis Juice Corp. for fresh juice processing. The sinusoidal wave form uses equipment comparable to that of a radio and is less difficult to generate. A square wave can deliver more energy per cycle because the sinusoidal only reaches maximum power for a split second. Although PEF is a non-thermal process, an increase in temperature occurs in the processing chamber.

A typical temperature change is about 30 °C for orange juice and less for apple juice. Processes typically operate at 35– 50 °C. Time of exposure depends on several factors, first of which is the chamber design of which there are two categories: flowing and non-flowing. Flowing processes include co-axial, parallel plate or co- field. In the co-field method, the electric field is cycled 1000 times a second through various treatment chambers, separated by ceramic or polymer insulators, while receiving multiple pulses. In nonflowing, the process is static and can be applied to solids. PEF imposes a strong electric field on pumpable foods for a very short time to kill vegetative cells. Critical field strengths of about 15000V/cm are used on foods, whereas at 35000V/cm, PEF is used as a disinfectant. Under PEF, cell membrane pores develop or enlarge, and can be reversible or irreversible. Pores affect membrane permeability by allowing external matter to enter, causing a loss of cellular content, thereby killing the cell. Perforation of cell membranes caused by PEF in fruit and vegetable cell walls can yield improved extraction of juice from cells.

Disadvantages that must be overcome in order to commercialize PEF:

- Scale up of the system in such a way that profitable production is possible.
- The presence of bubbles, which may lead to non-uniform treatment as well as operational and safety issues.
- Treatment of suspensions with solid particles, with a minimum risk of breakdown.
- Availability of commercial units.

If bubbles are present in the PEF treatment chamber, dielectric breakdown will occur. This happens because the spherical gas bubbles elongate, causing the ends to have up to a five times more intense electric field. The bubbles grow larger as the electric field overcomes the dielectric strength of the bubbles, causing partial discharge, and eventually connecting the two electrodes, causing a spark. Vacuum de-gassing and pressurized treatment during processing can minimize the presence of gas bubbles. Concerns must be addressed when considering PEF for the treatment of suspensions with particulates: include the potential for dielectric breakdown on the surface of particulates; uniform treatment distribution of the applied electric field; the manufacturing of a treatment chamber and feed pump system designed for particulates; control of heat induced by the process; and the particle size must be smaller than the gap in the treatment region to ensure proper processing.

Applications of Pulsed Electric Fields Technology

PEF is a continuous processing method, which is not suitable for solid food products that are not pump able. Pulsed electric fields technology has been successfully demonstrated for the pasteurization of foods such as juices, milk, yogurt, soups, and liquid eggs. Application of PEF processing is restricted to food products with no air bubbles and with low electrical conductivity. The maximum particle size in the liquid must be smaller than the gap of the treatment region in the chamber in order to ensure proper treatment. PEF processing has been successful in a variety of fruit juices with low viscosity and electrical conductivity such as orange, apple, and cranberry juice. Additionally, the color change in fruit juices (subject to prolonged storage) was reportedly less in juices treated by PEF, as in a recent study of PEF-treated orange juice stored at 4 °C for 112 days; there was less browning than thermally pasteurized juice, which was attributed to conversion of ascorbic acid to furfural. Considering the effectiveness of PEF treatment on liquid products, such as milk, fruit juices, liquid egg, and any other pumpable food products, extensive research has been done to implement the process at an industrial level. Flavor freshness, economic feasibility, improvements in functional and textural attributes and extended shelf life are some of the main points of interest besides achievement of microbiological safety of food products.

Ultrasound

Ultrasonic waves (energy generated by sound waves of 20,000 Hz or more) generate gas bubbles in liquid media, that produce a high temperature and pressure increase when they immediately burst. The mechanism of microbial killing is mainly due to thinning of cell membranes, localized heating and production of free radicals. These regions of pressure change cause cavitation to occur, and gas bubbles are formed in the medium. These bubbles have a larger surface area during the expansion cycle, which increases the diffusion of gas, causing the bubble to expand. A point is reached where the ultrasonic energy provided is not sufficient to retain the vapour phase in the bubble; therefore, rapid condensation occurs. The condensed molecules collide violently, creating shock waves. These shock waves create regions of very high temperature and pressure, reaching up to 5500 °C and 50,000 kPa. The pressure changes resulting from these implosions are the main bactericidal effect in ultra- sound. The hot zones can kill some bacteria, but they are very localized and do not affect a large enough area. The cavitation threshold of a medium (that is, the minimum oscillation of pressure that is required to produce cavitation) is determined by a number of factors. Among these are dissolved gas, hydrostatic pressure, specific heat of the liquid and the gas in the bubble, and the tensile strength of the liquid. Another extremely important variable is temperature, which is inversely proportional to cavitation threshold. The ultrasonic frequency used must be under 2.5 MHz, as cavitation will not occur above that level.

Mechanism and Effect of Ultrasonics

When sound energy passes to the medium resulting in a continuous wave-type motion, longitudinal waves will be generated with the result that the motion creates alternative compression and rarefaction of the medium particles. For food processing purposes it is important to address the generation of heat due to ultrasound applications and the related cavitation (implosion of gas bubbles) caused by a rapid change of heating to 550 °C and pressure increase to 50 Mpa. The temperature and pressure indicated are generated during a very short periods of time at the point were cavitation occurs with an order of temperature variation of 109 °C/s. Shock waves are generated due to cavitation, which are contributed to the ultrasound effect.

Among these are dissolved gas, hydrostatic pressure, specific heat of the liquid and the gas in the bubble, and the tensile strength of the liquid. Another extremely important variable is temperature, which is inversely proportional to cavitation threshold. The ultrasonic frequency used must be under 2.5 MHz, as cavitation will not occur above that level.

High Pressure Food Preservation

Novel food processing technologies aim to provide safe, high quality foods with desirable nutritional and physicochemical properties. More recently with the emergence of functional foods and nutriceuticals as well as with the increased use of minimal processing, health aspects of foods as well as gentle treatment concepts have been added to the already existing food processing requirements. This necessitates careful process design and results from intensive kinetic and nutritional evaluations for process development and monitoring.

The complex composition of most food and its large variety requires routes of processing which are both highly efficient in killing microorganisms and flexible enough to retain the desirable attributes of the product. Despite the extensive knowledge in food preservation by heat treatment and despite continued attempts to improve the quality of processed foods there is still a need for technologies that minimise the heat effects on desired quality attributes of foods. Even intelligent concepts like e.g. high-temperature-short-time processing fail if heat transfer and heat penetration is limited by intrinsic physical properties of the product. Because the thermal energy which is required to kill the contaminating microorganisms has to be conveyed across the product itself, the design of fast and uniform heating and cooling steps is one of the primary challenges of industrial heat preservation. Heat can be transferred by conduction, convection, and radiation. Most of the in use thermal processing equipment (except few applications of microwave, inductive or ohmic heating) use systems where the heat is transferred across interfaces driven by a temperature gradient. On the product side only the convective heat transport can be enhanced by external measures, i.e. by forced agitation. The transferable heat flow and the required time to warm up the centre of a solid or highly viscous product by solely heat conduction is determined by the thermal diffusivity of the material.

Non-thermal processing technologies make use of other physical principles of transmitting the energy to the target structures within the product. One of those emerging processes which could serve as an alternative method for food preservation is the use of high hydrostatic pressure.

Processing Technology

High pressure processing (HPP) of foodstuffs is used for the preservation and modification of foodstuffs. Thereby, foodstuffs are normally subjected for periods of a few seconds up to several minutes to hydrostatic pressures above 350 MPa. This treatment permits the inactivation of microorganisms and enzymes at low temperatures, whilst valuable low molecular constituents, such as vitamins, colours and flavourings, remain largely unaffected. The ability of hydrostatic pressures to inactivate microorganisms as well as to denature proteins was demonstrated about a hundred years ago. Over the last decades process development has progressed rapidly and high pressure treated foodstuffs have been marketed in Japan since 1990 and in Europe and the United States since 1996. Without doubt, the preservation of foods is by far the largest commercial application of high hydrostatic pressure related to biological systems, and the application has steadily increased during the past 10 years. At present, 128 industrial installations exist with volumes from 55 to 420 litre and a total annual production volume of more than 200,000 tons. Almost half of it is meat, meat products, seafood or fish. The rest are plant based products like vegetable preparations or different kinds of fruit juices.

Hydrostatic pressure is generated by increasing the free energy, e.g. by heating in closed systems or by mechanical volume reduction. Industrial high pressure installations are operated batch wise and can reach pressures up to 800 MPa. The pressure is then kept constant for a designated time which ensures the success of the process, typically from several seconds to several minutes. From chemical industry, where pressure is widely used to increase the reaction yield, the technology has been transferred to food and biotechnology. In biological systems pressure higher than 400 MPa can lead to a reversible and irreversible cleavage of intermolecular and intramolecular bonds. In this way structural changes in membranes as well as the inactivation of enzymes involved in vital biochemical reactions are the key targets of microbial kill by high pressure. The inactivation of virus is supposed to depend on the denaturation of capsid proteins essential for host cell attachment.

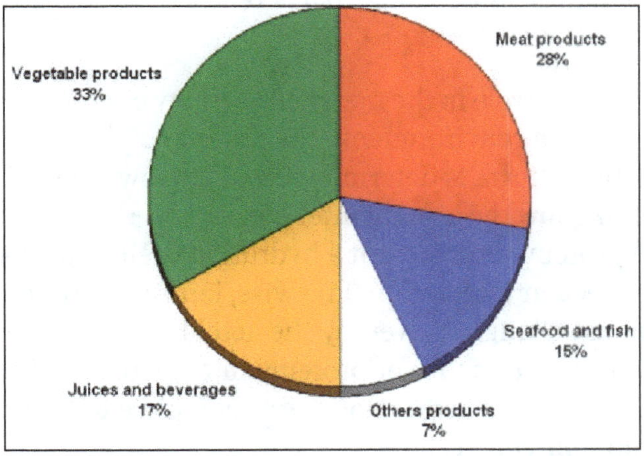

Utilization of HPP preservation in different segments of food industry.

HPP is a relatively young preservation technology. Compared to other methods which are commonplace in food industry like e.g. fermentation, drying or heating there is less experience in the specific features of HPP. The high pressure process itself is characterised by 3 parameters: Temperature T, pressure p, and pressure exposure time t. Compared to other processes like heat preservation which is based on two parameters only (T, t) the three parametric HPP offers a broad variability for process design. Table shows typical processing parameters for traditional and novel preservation treatments. In a qualitative approach, process efficiency is assessed in terms of the lethality of the treatment and its structural impact on the food matrix. Evidently, those treatments which are powerful in killing microbes have usually a strong destructive effect on the integrity of the food matrix with severe consequences on quality and consumer acceptance.

Traditional products are usually preserved by traditional technologies which, to a large extent, meet the expectations of the consumers of those foods. On the other side traditional preservation strategies fail or are not applicable when new product developments are based on innovative or uncommon ingredient compositions. In those situations nonthermal technologies like irradiation, pulsed electric fields or HPP came into the focus. The justification for applying novel preservation concepts should be: high safety margins, superior quality and reasonable costs.

Physical and Chemical Background

Within the last 20 years a considerable knowledge on the impact of high pressure on microbes, virus, food constituents and food structures has been accumulated and many practical applications of high pressure technology in food industry and biotechnology took advantage from the substantial advances in biochemistry and biophysics which led to an improved understanding of the mechanistical background.

In complex matrices like food the desired effect of e.g. microbial inactivation may also produce biochemical changes which may affect the product properties in a negative manner. The suitable selection of the processing parameters temperature, time and pressure can ensure that the processing goal is reached without extensive detrimental effects.

Proteins which play a major role in the metabolic activity of all living cells are extremely susceptible to changes in the environment. The stability of the protein's molecular configuration in its functional form is determined by a narrow band of parameter settings which impact mainly on how the protein interacts with the solvent. In the case of water, which is the natural protein environment a hydration shell is formed which itself is influencing the intramolecular interaction. Likewise, ionizable groups in lateral positions produce conformational changes driven by the actual proton concentration and ionic strength. The losses in functionality of proteins in response to those perturbations is hence related to intramolecular reorientations or complete unfolding leaving the polypeptide chain in a random-coiled state.

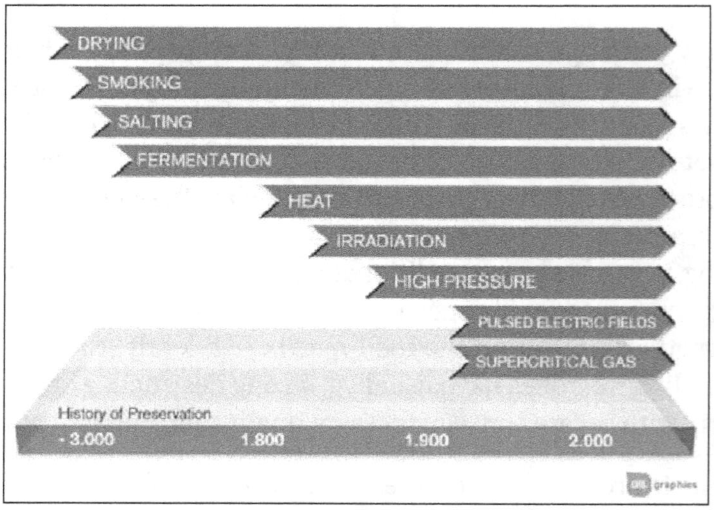

History of food preservation methods.

	Processing parameters	Processing intensity	Lethality	Structure impact	Cost [€/kg]
Drying	T, t	+	- -	++	5
Smoking	c, T, t	+	+	++	2
Salting	c, t	-	+	++	1
Fermentation	b, T, t	- - -	+	+	2
Heat	T, t	+++	+++	+++	0.05
Irradiation	$w	++	+++	++	0.3
High pressure	p, T, t	+	++	+	0.3

Pulsed electric fields	E, $w	++	++	+	0.1
Supercritical gas	c, p, T, t	+	+	+	1

b microbial growth parameters, c chemical efficiency parameters, E electric field strength, $w specific total energy input, p pressure, T temperature, t pressure exposure time,

In many cases, the 'functional' state is considered as the native state whereas the 'unfunctional' state is referred to as the denatured state no matter which particular molecular structure they form. Nevertheless, there are means to discriminate between both states and thermodynamic potential functions like Gibbs free energy are particularly useful when the transition from the native to the unfolded configuration is under consideration. In those situations the difference in Gibbs free energy ΔG is reduced to zero which occurs only at d settings of the relevant physical and chemical parameters: $\Delta G = \Delta G_0$? $f(T, p$, pH, co-solvents). If all parameters apart from T and p are fixed, the slope of the phase boundary is described by the equation of Clausius - Clapeyron:

$$\frac{dp}{dT} = \frac{\Delta S}{\Delta V}.$$

In those situations, the transition is accompanied by an exchange of latent heat with the environment. Generally, the existence of a phase transition is based on an enthalpic and an entropic contribution to the free energy function:

$$d(\Delta G) = \Delta V dp - \Delta S dT.$$

An equation (Eyring equation) has been derived from the transition-state theory, relating pressure and the rate constant k of reactions under pressure using the activation volume $\Delta V^{\#}$ as a parameter:

$$\left(\frac{\partial \ln k}{\partial p}\right) = -\frac{\Delta V^{\#}}{RT}$$

The functional associations of pressure, temperature and reaction time are best presented by means of pressure–temperature diagrams (pTdiagrams), which show pressure-temperature combinations that will lead to a desired reaction (e.g. inactivation) rate constant. Thus, a database software was particularly designed to enable the user to call up pressure–temperature function equations for a number of microorganisms, enzymes and food constituents and to present them in pT-diagrams for predetermined treatment times or as kinetics under predetermined p-T conditions (Buckow and Heinz.

HPP Inactivation of Vegetative Bacteria

The main application of HPP in the food industry is for the extension of shelf-life or for the elimination of microbial pathogens. The viability of vegetative microorganisms may be affected by inducing structural changes at the cell membrane or by the inactivation of enzyme systems which are responsible for the control of the metabolic actions. Typically, significant inactivation of vegetative bacteria, yeasts and moulds viruses can be observed within minutes at room temperature and pressures higher 300 MPa. However, increasing the pressure to 700 MPa or higher most inactivation reactions are strongly accelerated.

So far there only a few studies reporting inactivation kinetics of vegetative microorganisms over a wide range of pressure–temperature combinations. Figure exemplarily shows pressure–temperature combinations that lead to 5 log reduction of several pathogenic and spoilage organisms within 5 min of treatment. It is generally accepted that pressure and temperature act synergistically on the destruction of vegetative bacteria in the high temperature domain, which is indicated by the left bend of isorate curves in. However, L. casei seems to represent an exception as counter-effects of pressure and temperature has been observed for its inactivation in 10 mm HEPES buffer (pH 5.3). As can be seen from, pressure stability of microorganisms often appears to be maximal at 20–40 °C whereas stability is decreased at lower temperatures. This might be explained by the increase of water and cell cytoplasm compressibility with decreasing temperature and thus, an increased transfer of mechanical energy to the microbial cell. Assuming microbial cell death is initiated at a certain threshold of mechanical energy transferred into the cellular system, at low temperatures this lethal threshold is achieved at lower pressures than the pressure needed at higher temperatures.

HPP Inactivation of Bacterial Spores

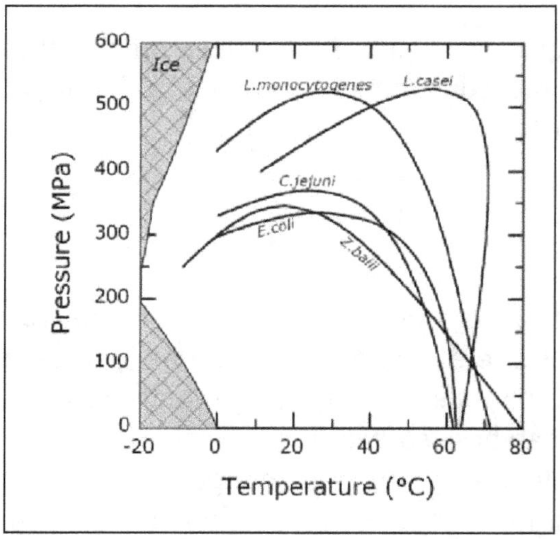

Pressure–temperature isorate diagram for 5 log inactivation of C. jejuni, E. coli and L. casei and Z. balii after 5 min isothermal/isobaric treatment.

Bacterial spores are not by themselves an hazard to the food industry. It is the eventual germination, outgrowth, and proliferation of the organism which results in toxification or spoilage of food during the post-processing storage. Bacterial endospores, as compared to vegetative cells, display a considerably higher resistance to temperature and high pressure. To cope with this potential hazard, three strategies are in use to minimise the risk of spore contamination:

- Full inactivation in one step by severe temperature conditions or suitable pressure temperature combinations.
- Germination spores by temperature and pressure and inactivate them in a subsequent temperature or pressure/temperature treatment (milder than strategy 1),
- Injury of spores by temperature or pressure/ temperature treatment (milder than strategy 1) and prevent germination or outgrowth in the food by matrix inherent hurdles.

At present the database only considers those published results that reach spore inactivation by a one step process (strategy 1). Spores of Clostridium botulinum and Bacillus species are the key bacteria for the safety or the spoilage of low acid (heat treated) preserved goods. These spores have shown remarkable tolerance to pressures above 1,000 MPa at room temperature. On the other hand, many other bacterial endospores, which are relevant to food are inactivated at pressures 600 MPa or greater in combination with initial temperature above 60 °C. Often the required inactivation temperature and time is lowered by combination with pressure as indicated in the pressure-temperature plain of figure for a number of bacterial spores.

HPP Inactivation of Virus

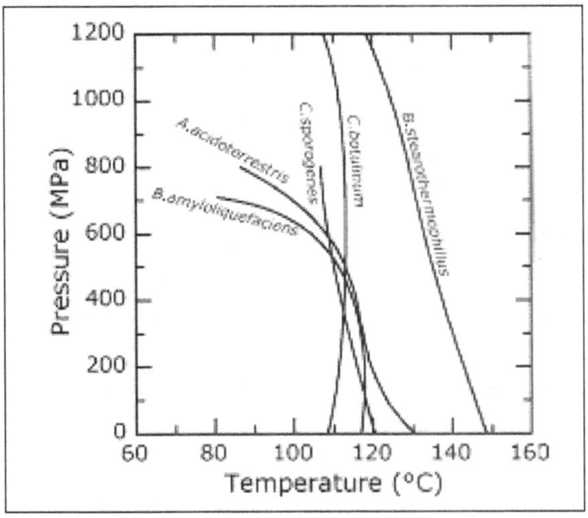

Pressure–temperature isorate diagram for 5 log inactivation of A. acidoterrestris, B. amyloliquefaciens after 5 min isothermal/isobaric treatment.

Viruses, regardless of their type of envelope, show a wide range of sensitivities in response to high hydrostatic pressure. It has been suggested that virus inactivation by high pressures is due to denaturation the capsid proteins essential for host cell attachment to initiate infection but leaves the actual capsid and RNA intact. For protein unfolding it has already been stated that high pressure can not be seen independently from the temperature at which the treatment is performed. Thus, it is not surprising that also the pressure stability of viruses is greatly affected by the process temperature. However, in contrast to proteins, pressure induced stabilization of viruses towards heat inactivation is not a general phenomenon, but has been observed in isolated cases. On the other hand, a number of reports have indicated that the dissociation and denaturation of proteins and viruses by pressure is promoted by low temperatures. Such behaviour is exemplarily shown in figure for selected viruses and might be explained by an increased exposure of nonpolar protein side chains to water at low temperatures. This leads to enhanced interactions of nonpolar groups causing partial denaturation of proteins at elevated pressures.

HPP Modification of Food Constituents (Starch, Protein and Fat)

The primary structure of low molecular weight molecules such as vitamins, peptides, lipids, and saccharides is rarely affected by high pressure because of the very low compressibility of covalent bonds at pressures <2,000 MPa. On the other hand, certain macromolecules, such as starches, can change their native structure during HPP, in a manner analogous to thermal treatments. For example, starch granule solutions can form very smooth starch pastes, which can be used to replace fat in reduced energy foods. Starch granules solutions can form a week gel due to pressure induced swelling of the granule. Therefore, the occurrence of intermediate degradation levels of the lamellar crystalline regions of the starch granule can be anticipated, which is a possible reason for the significant difference, e.g. in viscosity between starch gels formed at different pressure/temperature conditions.

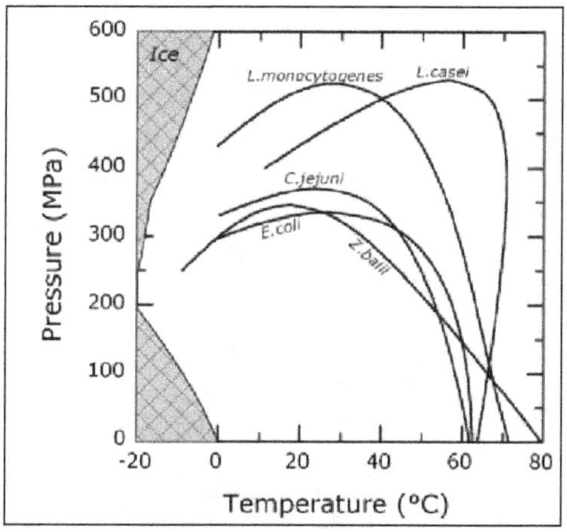

Pressure–temperature isorate diagram for 5 log inactivation of Avian influenza virus (in chicken meat slurry), Feline calici virus (in eagle medium), Coxsackie virus and Rota virus after 1 min isothermal/isobaric treatment.

The pressure range in which gelatinization occurs is specific for each starch and is partly dependent on its crystalline structure, e. g. B-type starches are more resistant to pressure than A- and C-type starches and the proportion of amylose and amylopectin. Usually the extent of gelatinization reached depends on pressure level, treatment temperature and processing time. The pT diagram of figure shows a compilation of phase transition lines (native/gelatinized) of different starch-water suspensions. It is evident that the gelatinization temperature of starch granules is decreased when the pressure exceeds a specific threshold level. However, the pressure effect is by far smaller when the treatment temperature is approaching the gelatinization temperature at atmospheric pressure. Wheat starch is different in that sense because the temperature required to irreversibly change the native granule structure is already decreased markedly when pressure is increases by 50 MPa.

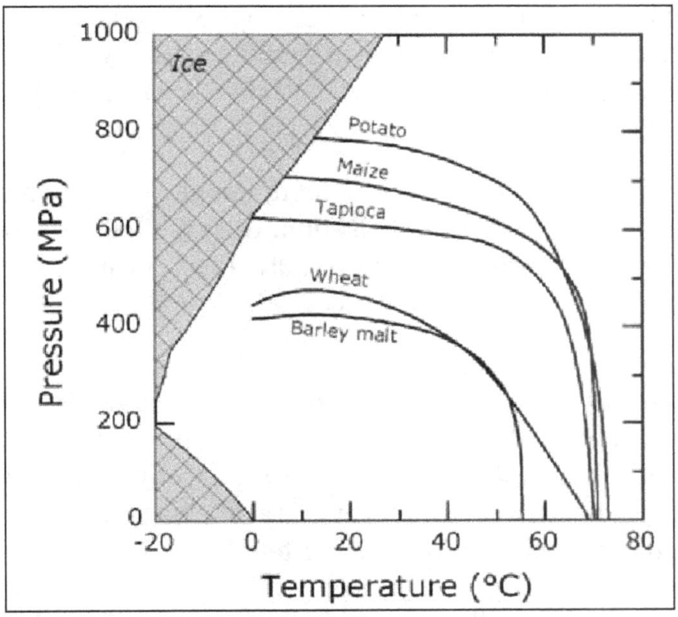

Pressure–temperature phase diagram for complete gelatinization of starch granules from barley malt, maize, potato, rice, tapioca and wheat after approximately 15 min isothermal/isobaric treatment.

Legislative Aspects

Prior to introducing novel foods to the market food companies need to get an approval that those products are in compliance with the food law. With regard to the manufacturing process the question may arise whether a technology that can be considered as "novel", is necessarily producing "novel food" within the meaning of the law. The "Novel Foods Regulation" (Regulation (EC) No 258/97) defines novel food as a food, that does not have a significant history of consumption within the European Union (EU) before the 15th of May, 1997. Such foods are subject to a pre-market safety assessment, before a decision is made on EU-wide authorisation.

The original intention of the Novel Foods Regulation was to introduce a legal framework for foods and food ingredients containing or consisting of genetically modified organisms (so called GMO). For the first time, this legal framework provided the possibility to allow GMO in food stuffs. Because of the fact that GMO were (and are) politically highly disputed, the European legislator tried to "hide" this new approach of allowing GMO and used the term of "Novel Foods", regulating not only GMO, but also other kinds of novel foods like — for example — foods consisting of micro-organisms, fungi or algae. Nowadays, GMO are not subject of the Novel Foods Regulation any more. Today, GMO form a separate legal category and they are regulated by separate provisions.

What concerns processing aspects, the legislations defines Novel Foods as follows:

Foods and food ingredients to which has been applied a production process not currently used, where that process gives rise to significant changes in the composition or structure of the foods or food ingredients which affect their nutritional value, metabolism or level of undesirable substances.

If a food falls under the definition of novel food, the person responsible for placing it on the market has to apply for an authorisation. Only if a food was commercialised in at least one Member State before the 15th of May, 1997, it can be marketed elsewhere in the EU under the "principle of mutual recognition", and the Novel Foods Regulation does not apply.

In order to be granted the authorisation, the applicant submits a request to the Member State in which the product is to be placed on the market for the first time. The request shall contain the necessary information, including a copy of the studies which have been carried out and any other material which is available to demonstrate that the food complies with the demanded criteria. These criteria are:

Foods and food ingredients falling within the scope of this Regulation must not:

- Present a danger for the consumer,
- Mislead the consumer,
- Differ from foods or food ingredients which they are intended to replace to such an extent that their normal consumption would be nutritionally disadvantageous for the consumer.

Furthermore, the applicant has to provide an appropriate proposal for the presentation and labelling of the food.

Once the application has been accepted, the Member State has 90 days to produce an initial opinion. This opinion is then circulated in all EU Member States, who are then given a further 60 days period to comment or make a reasoned objection. If there are no objections, the novel food will be authorised (or rejected) at the end of the 60 days in line with the initial opinion. Otherwise, a decision on the authorisation will be taken by a vote among Member States at the Standing Committee on the Food Chain and Animal Health. If necessary, the European Food Safety Authority will first be asked for its opinion on any outstanding safety questions.

The Novel Foods Regulation includes a simplified procedure for marketing certain types of novel food or novel food ingredient in the EU, if it is considered "substantially equivalent" to an existing food or food ingredient that is already marketed within the EU. In these cases, the company can submit a notification to the European Commission after obtaining an opinion on equivalence from an EU Member State.

In the UK for example, it is the Food Standards Agency who is the Competent Authority. If a company wants to benefit from the simplified procedure, it must provide the Competent Authority with specific data when requesting such an opinion. The company's application dossier should show how the novel food or novel food ingredient may be substantially equivalent to an existing food or food ingredient as regards to its: (a) composition (such as the source organism and preparation method), (b) nutritional value, (c) metabolism, (d) intended use (such as a food ingredient or supplement) and (e) level of undesirable substances (such as contaminants, mycotoxins and allergens).

Non-thermal Plasma

Atmospheric pressure plasma (APP) is an emerging non thermal technology for the improvement of food safety. Non-thermal plasma (NTP) is a neutral ionized gas that comprises highly reactive spices including, positive ions, negative ions, free radicals, electrons, excited or non excited molecules and photons at or near room temperature. NTP can be generated at atmospheric pressure that makes it more applicable. Moreover, it could be employed in inactivation of microorganisms on the surface of fresh and processed foods. However, for the reason that there are few studies on the application of this technology in real food systems, the effects of non-thermal plasma on nutritional and chemical properties of food is not known well. Furthermore, the studies which explore the safety and cost aspects of this technology could help it become widespread in food industry.

Food safety is a major concern for food industry, regulatory agencies as well as consumers. Food-borne pathogens and spoilage microorganisms are problematic microbes in food industry due to its significant public health risks and economic impact. There are a lot of sterilization method to eliminate these microorganisms. Some of these methods rely on lethal heat treatment such as steam pasteurization, autoclaving, ohmic heating, etc. Thermal technologies have side-effects on nutritional, sensory and functional properties of treated foods so alternative non thermal pasteurization methods such as high hydrostatic pressure, pulsed electric field, oscillating magnetic field, ionizing irradiation and high power ultrasound have been developed and studied in recent years. These processes retain quality of foods better than conventional methods but they are cost effective, required specialized equipment and trained personnel and also consumer acceptances and safety of these processes should be considered.

Non thermal plasma is a new discipline in food processing. Plasma is electrically energized matter in a gaseous state that can be generated by electrical discharge. Electrical discharges in atmospheric pressure and low temperature make this process practical, inexpensive and suitable for decontamination of products which heat is not desirable for.

Fundaments of Plasma

Plasma Definition, Generation and Classification

Plasma is ionized gas, that consists of a large number of different species such as electrons, positive and negative ions, free radicals, gas atoms, molecules in the ground or excited state and quanta of electromagnetic radiation (photons). It is considered to be the forth state of matter in the world. It can be generated in the large range of temperature and pressure by means of coupling energy to gaseous medium. This energy can be mechanical, thermal, nuclear, radian or carried by an electric current. These energies dissociate the gaseous molecules into collection of ions, electrons, charge – neutral gas molecules and other species. Depending on the type of energy supply and amount of energy transferred to the plasma, density and temperature of the electrons are changed. These lead Plasma to be distinguished into two groups, high temperature plasma and low temperature plasma.

High temperature plasma implies that electron, ions and neutral species are in a thermal equilibrium state. Low temperature plasma is subdivided to thermal plasma, also called local thermodynamic equilibrium plasmas (LTE) and non thermal plasma (NTP), also called non-local thermodynamic equilibrium plasmas (non-LTE). An equilibrium or near equality between electrons, ions and neutrals is the main characterization of thermal plasmas (TP). Frequently employed thermal plasma generating devices are those produced by plasma torches, and microwave devices. In generation of cold plasma most of the coupled electrical energy is channeled to electron component instead of heating entire gas stream so the temperature of heavy particle remains near the room temperature, these characteristics make it suitable to be used in processes which high temperature is not desirable.

Table: Classification of plasma.

Plasma	Properties	Example
High temperature plasma (Equilibrium plasma)	$Te \approx Ti \approx Tg$, $Tp = 10^6 - 10^8$ K	Laser fusion plasma
	Low temperature plasma	
Thermal plasma (Quasi-equilibrium plasma)	$Te = Tg = T \leq 2 \times 10^4$ K $n_e \geq 10^{20}$ m^{-3}	Arc plasma
Non-thermal plasma (Non-equilibrium plasma)	$Te >> Ti \approx Tg = 300 - 10^3$ K $n_e \approx 10^{10}$ m^{-3}	Glow discharges

Atmospheric Pressure Plasmas (APP)

Low pressure glow discharge plasmas are of great interest in microelectronic industries but their vacuum equipment limits their application. Therefore one of the recent challenges was developing new plasma sources that can operate at or near 1 atmospheric pressure. Power sources of atmospheric pressure plasma generation can be microwave, RF (radio frequency), pulsed, AC (alternating current) or DC (direct current).

Devices that have been used for plasma generation are the corona discharges, micro hollow cathode discharges, gliding arc discharge, one atmospheric uniform glow discharge, dielectric plasma needle barrier discharge (DBD), atmospheric pressure plasma jet (APPJ).

Among all. DBD and APPJ are commonly used in industrial application like lightening, surface modification, etching and deposition.

Microbial Inactivation Mechanism of Plasma

The use of plasma as a sterilization method was first patented in 1968 and the plasma made from oxygen was first applied in 1989 after that considerable researches have been done on mechanisms of microbial inactivation by plasma agent. Interacting plasma agent with biological mater contributed to lethal action. Plasma treatment can effectively inactivate a wide range microorganisms including spores and viruses. Effect of plasma on different microorganisms can be completely selective, meaning that it can damage pathogenic microorganisms without damaging host or it can activate different pathways in different organisms.

The reactive spices in plasma have been caused the oxidative effects on the outer surface of microbial cells. Nitrogen and oxygen gas plasma are good sources of reactive oxygen- based and nitrogen-based spices such as O, O_2, O_3, O_H, NO, NO_2. Chemical rate constant of atomic oxygen for oxidation at room temperature is so higher than molecular oxygen. These act on the doubled bond of unsaturated fatty acid of membrane cell, thereby disturbing the transport of bimolecules across it.

In spite of oxidation of the lipids, amino acids and nucleic acids of cells and spores are vulnerable to the action of these spices and oxidation of them cause changes that lead to microbial death or injury.

In addition to reactive spices, UV photons can modify the DNA of the microorganisms and as a result disturb cell replication. Many studies have found that reactive spices had the most important role in inactivation of microorganisms and the role of UV photons in plasma was minor, but these studies demonstrated that more researches need to be done over the role of UV photons in plasma.

Contribution of each of the above mentioned mechanisms in inactivation microorganism depends on plasma characteristics and to the type of microorganisms. The former includes device set up (reactor geometry), voltage, gas pressure, gas composition, water content in the gas, and distance of the microorganism from the discharge glow, where the latter takes account of Gram-positive, Gram negative, spores and other types. To cite an example, compared the destructive efficiency of different gas composition and temperatures on Bossilus spp. Spores. They found that oxygen based plasma is more efficient than pure argon plasma. The other criterion that was considered is the direct exposure or remote exposure of substances from the plasma sources, The recent findings demonstrated that if the substrate which is sterelizied to be in indirect contact (remote exposure) with the plasma, the quantum of heat transmitted to a sample is reduced and many of the short-lived reactive spices do not reach the sample so the treatment cannot be efficient in microorganism inactivation.

Potential Application in Food

NTP has been applied in the food industry including decontamination of raw agricultural products (Golden Delicious apple, lettuce, almond, mangoes, and melon), egg surface and real food system (cooked meat, cheese,). In one study on E. coli 12955 a non-pathogenic surrogate for Salmonella spp. inoculated onto almonds, reported a reduction of more than 4 log CFU/ml after 30 s treatment at 30 kV and 2000 Hz. Similarly, Niemira and Site reported the reduction of Salmonella and E. coli 0157:H7 that inoculated onto apple surfaces for 2.9 – 3.7 and 3.4 – 3.6 log CFU/ml respectively, and they reported the highest flow rate of the air (40 liters/min) in discharge medium would be the most effective In this study cold plasma was generated in a gliding arc . studied decontamination effect of NTP generated by an AC voltage (variable 12 – 16 kV) on pericurp of melon and mangoes that inoculated by Sacchearomyces cerevisae, Pantoea agglomerans, Gluconacetobacter liquefacien and Ecoli. It was observed that S. cervisie was the most resistant. P. agglomerans and G. liquefaciens were reduced below the detection limit (corresponding to 3 log) after only 2.5 s on both fruits, whereas E. coli required 5 s to reach the same level of inactivation.

Salmonella spp. has been reported largely as a potential hazard for egg consumers. Decontamination of egg surface using barrier discharge plasma was studied by Ragni et al. The results showed that maximum reduction of 2.2–2.5 log CFU/eggshell in Salmonella enteritidis levels achieved after 60 – 90 min treatment at 35% RH. Further, higher RH lead to higher effectiveness of treatment, at 65% RH, reduction of 3.8 and 4.5 log CFU/eggshell were achieved after 90 min of exposure. Similar result was observed for Salmonella typhimurium, with an overall reduction of 3.5 log CFU/eggshell.

Montenegro employed direct current corona discharges for reduction of E. coli O157:H7 in apple juice. After 40 s treatment at a frequency of less than 100 Hz with 4000 pulses of 9000 V peak voltage, the number of cell reduction was more than 5 log CFU/g.

One of the factors that influence the effectiveness of NTP is the type of food which is being treated. Song et al. investigated the influence of this factor. In this study sliced cheese and ham inoculated by 3- strain cocktail of Listeria monocytogenes (ATCC 19114, 19115, and 19111, LMC) and exposure to barrier discharged plasma (75, 100, 125, and 150 W) for 60, 90 and 120s. Microbial log reduction increased with increases of input power and exposure time. The results indicated that, reduction after 120s at 75, 100, 125, 150 W ranged from 1.7 to 8 log CFU/g in sliced cheese and those in sliced ham were 0.25 to 1.73 log CFU/g. These result confirmed that type of food has strong effect on inactivation of LMC.

The Limitations of APP process for food sterilization are that, in treatment of bulky and irregularly shaped food, restricted volume and size of the food should be considered and also microbial inactivation occur on the surface of the food being treated since plasma reactive species are limited to penetrate into foods.

Future Prospect of Atmospheric Pressure Plasma

Atmospheric pressure plasma is proved to have specific potential for treatment of foods. Combining APP with other non thermal processes could be a possible future breakthrough in this field. In this case, synergistic effects may be more considerable however scaling up this technology remains a challenge to be solved. One of the dark aspects of experimental work on APP is that, treatment must be proven not to have negative impact on the organoleptic and nutritional properties of food; nevertheless, there has been limited investigation on this aspect of treatment. Hence, it is a necessary for further studies to specify the extent in which APP affect the chemical and nutritional properties of foods and its shelf life. In addition, risk assessment of toxic residues should be carried out in future works. The last but not the least, evaluation of the projected cost of treatment and safety of applied gas for scaling up this technology in food industry should be.

Controlling Water Activity

Food preservation involves the action taken to maintain the desired properties or nature of foods, within a time frame, so that it remains safe and pleasant to consume. A stable food product can be developed by applying different processing techniques and by keeping it in appropriate conditions. Food stability determination from a scientific basis rather than empiricism is a challenge to food scientists and engineers.

Water is an important basic element in foods. For a long time, the industry has known how important it is to check free water. The water activity (a_w) measurement forms the basis of this and provides important information about the quality of a product. Finally it provides information regarding the possibility of microbiological growth on the surface. Only with this conclusions can be made about the stability and durability of a sample.

The relative equilibrium humidity of a product, which is ascertained through its partial pressure of water vapor on the surface, depends on the following factors: Chemical compound; Temperature; Water content; Storage environment (T/RH); Absolute pressure; Packing.

Free water in products is jointly responsible for the growth of undesirable organism such as bacteria or fungi, which produce "toxins" or other harmful substances. Bat also chemical/biochemical reactions (e.g. the Maillard reaction) increasingly take place and possibly change the following factors of a product: Microbiological stability, Chemical stability, Content of proteins and vitamins, Color, taste and nutritional value, Stability of the compound and durability, Storage and packing, Solubility and texture.

Water activity is a fundamental property of aqueous solutions, and by definition is the ratio of the vapor pressure of the water in the substrate (p) to that of pure water at the same temperature (p_o):

$$a_w = \frac{p}{p_o},$$

Water activity is a measure of how efficiently the water present can take part in a chemical (physical) reaction. If half the water is so tightly bound to a protein molecule that it could not take part in a hydrolysis reaction the overall water activity would be reduced. Water activity (a_w) is defined as where p and P_o are the partial pressures of water above the food and a pure solution under identical conditions respectively. The tightly bound water has no tendency to escape from a food as a vapor and therefore exerts no partial pressure and has an effective water activity of zero. Water activity is clearly a function of composition but is also a function of temperature. The a_w is related to the boiling and freezing points, equilibrium relative humidity (ERH; see above equation), and osmotic pressure. Water activity ranges from zero (water absent) to 1.0 (pure water). For an ideal solution a_w is independent of temperature, and in actual practice, a_w of a given solution varies only slightly with temperature within the range of temperature permitting microbial growth. The relationship between water potential and water activity is given by the next equation, where the value of k depends on temperature and is, for example, 1.37 at 25 °C and 1.35 at 20 °C.

$$\Psi\left(M_{pa}\right) = K = \text{In}\, a_w$$

Not only is the availability of water in the surrounding liquid phase of importance to fungi, but the water content of the adjacent gas phase. The water content of the atmosphere is expressed in terms of relative humidity, the ratio of the water vapor pressure of the gas phase being considered to that of a saturated atmosphere at the same temperature. It is hence the same ratio as water activity but expressed as a percentage.

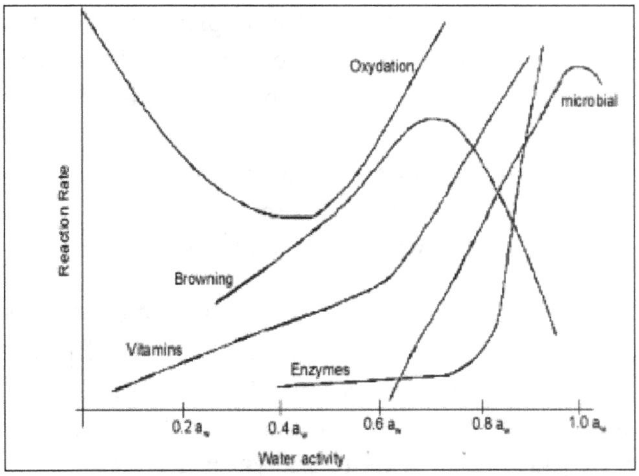

The variation of oxidation, browning, enzyme activity, vitamin inactivation and microbial activity.

Water activity (a_w) indicates the availability of water's medium for chemical reactions, biochemical transfer or exchange through a semi permeable membrane.

Result's Law is very accurate within the range of 0.95 to 1.0 water activity, whereas the Norrish equation is reasonably accurate down to about 0.55. Result's Law applies to solutions of small molecular weight compounds for which the calculated $a_w > 0.95$.

$$a_W = An_{H2O} / n_{H2O} + n_{solute}$$

where: A- activity coefficient; n_{H2O}- moles of water in solution (assumed to be 1 for solutions with $a_w > 0.95$); n_{solute} - moles of solute.

The Norrish equation is used for solutions, and is valid when the molecular weight and the Norrish k value are known. For a single solute, the Norrish equation reduces to:

$$a_w = X_W \times \exp\left(-k_i \times X_i^2\right)$$

where: aw = water activity;

X_W = mole fraction of water;

X_i = mole fraction of solids in ingredient i;

k_i = Norrish constant for ingredient i.

Present work evaluated the ability of Norrish's equation to model the water activity of solutions of sugars.

Water Activity (a_w) Concept

The concept of aw has been very useful in food preservation and on that basis many processes could be successfully adapted and new products designed. Water has been called the universal solvent as it is a requirement for growth, metabolism, and support of many chemical reactions occurring in food products. Free water in fruit or vegetables is the water available for chemical reactions, to support microbial growth and to act as a transporting medium for compounds. In the bound state water is not a available to participate in these reactions as it is bound by water soluble compounds such as sugar, salt gums, etc. (osmotic binding), and by the surface effect of the substrate matrix binding).

These water-blinding effects reduce the vapor pressure of the food substrate according to Raoult's Law. Comparing this vapor pressure with that of pure water (at the same temperature) results in ratio called water activity (a_w). Pure water has an a_w of 1, one molar solution of sugar – 0.98, and one molar solution of sodium chloride -0.9669. A saturated solution of sodium chloride has a water activity of 0.755. This same NaCl solution in a closed container will develop an equilibrium relative humidity (ERH) in a head space of 75.5%. A relationship therefore exists between ERH and aw since both are based on vapor pressure:

$$a_w = \frac{ERH}{100}$$

The ERH of a food product is defined as the relative humidity of the air surrounding the food at which the product neither gains nor loses its natural moisture and is in equilibrium with the environment. The definition of moisture conditions in which pathogenic or spoilage microorganisms cannot grow is of paramount importance to food preservation. It is well known that each microorganism has a crystal aw below which growth cannot occur. For instance, pathogenic microorganisms cannot grow at $a_w < 0.62$. The socalled intermediate moisture foods (IMF) have aw values in the range of 0.65 - 0.90.

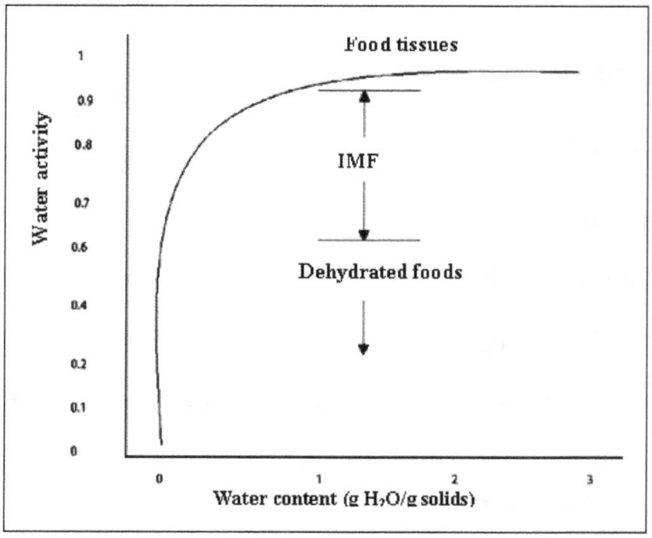

Typical equilibrium of water content vs. water activity in foods.

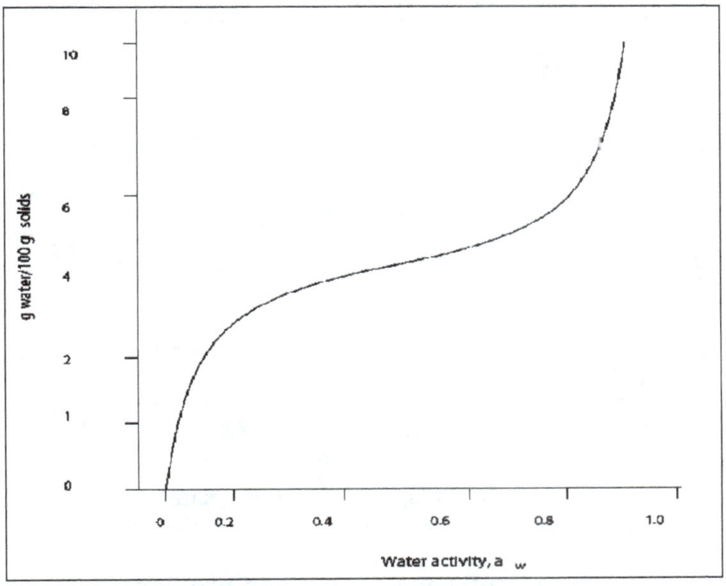

Equilibrium of water activity vs. moisture content, typical in foods Lower region of isotherm.

With a_w at 0.3, the product is most stable with respect to lipid oxidation, non-enzymatic browning, enzyme activity, and of course, the various microbial parameters. As aw increases toward the right, the probability of the food product deteriorating increases.

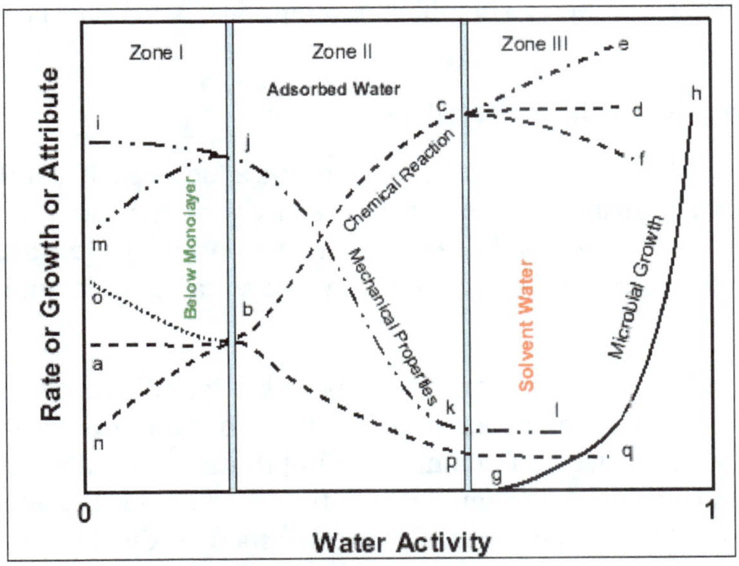

Stability diagram based on the water activity concepts.

where: *gh* = Microbial growth trend; *oa, ab, nb* = chemical reaction trends below BET-monolayer; *bc*, bp = chemical reaction trends in the adsorbed water; *ce, cd, cf, pq* = chemical reaction trends in the solvent water region; *ij, mj* = mechanical properties trends below BET-monolayer; *jk* = mechanical properties trend in the adsorbed water region; *ki* = mechanical properties trend in the solvent water region.

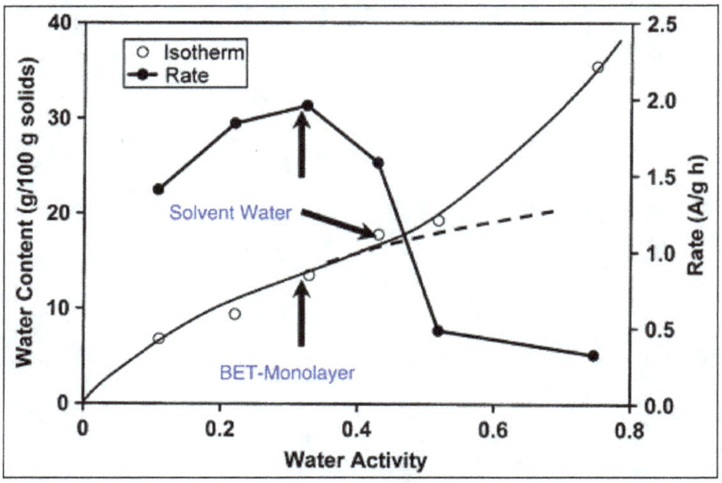

Enzymatic browning rate at 70°C in gelatinized starch medium. Indicate water activity at BET-monolayer.

In general the rule of water activity concept is: Food products are most stable at their "BET monolayer moisture" content or "BET-monolayer water activity" and unstable above or below BET monolayer. However, experimental evidence showed that optimal moisture for stability was in the multilayer adsorption region. In many other instances it has been shown that optimal water content for stability is not exactly the BET-monolayer. The reason for this variation is due to the fact that the BET theory of adsorption was developed based on many simplified assumptions, which are not realistic when food is considered.

Water Activity and Shelf Stability

Water activity, unlike water content, can determine a food's shelf stability. It can predict which microorganisms will be potential sources of spoilage and infection (the difference between bacterial pathogens and fungal physiology, or, a_w of 0.91 versus that of 0.70). The water activity of a food is instrumental in maintaining its chemical stability.

Consider that water activity is partially responsible for minimizing non-enzymatic browning reactions and spontaneous autocatalytic lipid oxidization reactions; prolonging the activity of enzymes and vitamins; and optimizing the physical properties of products such as moisture migration, texture, flavor, odor and shelf life. Not bad for a little relative humidity measurement. Every retail food establishment needs to know what will happen to their products as they sit on the shelf, even under ideal conditions of temperature and humidity. Shelf stability means the product 'won't get moldy, but it also affects the foods texture, moisture migration and caking and clumping.

Stability and food security depends on water activity and pH in the food environment. The water activity is higher with the products are perishable. Bat, even at low pH values and low aw, certain yeast and mould species that can tolerate high solute concentrations might pose a risk to the stability of Intermediate Moisture Foods (IMF). Currently, consumers prefer foods (vegetables and fruits) processed at least. Therefore, safety considerations are addressed seriously by food microbiology.

There are different approaches to conservation and stability of fresh fruit products. Commercial, minimally processed fruits are fresh (with high moisture), and are prepared for convenient consumption and distribution to the consumer in a fresh-like state. Minimum processing includes preparation procedures such as washing, peeling, cutting, packing, etc., after which the fruit product is usually placed in refrigerated storage where its stability varies depending on the type of product, processing, and storage conditions. However, product stability without refrigeration is an important issue not only in developing countries but in industrialized countries as well. The principle used by Leistner for shelf-stable high moisture meats ($a_w > 0.90$), where only mild heat treatment is used and the product still exhibits a long shelf life without refrigeration, can be applied to other foodstuffs. Fruits would be a good choice. Leistner states that for industrialized countries, production of shelf-stable products (SSP) is more attractive than IMF because the required a_w for SSP is not as low and less humectants and less drying of the product is necessary.

This Manuel presents information related to the processing of fruit and vegetables by combined methods. Information concerning the trade and production of fruits and vegetables in different countries is provided, as well as information on the processing of fruit and vegetable products. The combination of factors such as water activity (a_w), pH, redox potentials, temperature, and incorporation of additives in preserving fruits and vegetables is important, and all play a crucial role in improving the shelf life of fresh and processed commodities. During the last decade, minimally processed high moisture fruits (HMFP) which are ambient stable (with $a_w > 0.93$) have been developed in Latin American countries, under the leader of Argentina, Mexico and Venezuela.

The methodology employed was based on combinations of mild heat treatments, such as blanching for 1-3 minutes with saturated steam, slightly reducing the a_w (0.98-0.93) by addition of glucose or sucrose lowering the pH (4.1 -3.0) by addition of citric or phosphoric acid, and adding antimicrobials (1000ppm of potassium sorbate or sodium benzonate, as well as 150ppm of sodium sulphite or sodium bisulphite) to the product syrup. During storage of HMFP, the sorbate and sulphite levers decreased, as well as a_w levels, due to hydrolysis of glucose.

A variety of alternative method to preserve fruits and vegetables can be used in rural areas such as fermentation sun drying, osmotic dehydration, and refrigeration. Fruit and vegetables can be preprocessed via blanching to eliminate enzymes and microorganisms. Over the last decade, have been developed innovative technologies for obtaining shelf-stable "high moisture fruit products" (HMFP) storable for 3-8 months without refrigeration.

These new technologies are based on a combination of inhibiting factors to combat the deleterious effects of microorganisms in fruits, including additional factors to diminish major quality loss in reactions rates. Slight reduction of water activity (a_w 0.94-0.98), control of pH (pH 3.0-4.1), mild heat treatment, addition of preservatives (concentrations £ 1.500ppm), and anti browning additives were the factors selected to formulate the preservation procedure. These techniques were preceded by the pioneer work of on the combined effects of several factors applied to meat products - named “hurdle” technology.

Prediction of Water Activity in Practical Applications

Water activity a_w can be influenced in at least three ways during SX the preparation of dried, intermediate and high moisture foods:

- Water can be removed by a dehydration, evaporation or concentration process.
- Additional solute can be added. Theirs penetration of solute can be performed by moist infusion or by dry infusion. Moist infusion consists in soiling the food pieces in a water/solute solution of lower a_w while dry infusion involves direct mixing of food pieces and solute in required proportions. When water/rich solid products, such as fruit and vegetables, are subjected to moist or dry infusion, three flows arise:
    - A water outflow, from product to the environment.
    - A solute flow, from the environment to product.
    - An outflow of the product’ sown solutes.

This process is called’ osmotic dehydration’ and allow the infusion of not only the solute used to control a_w bat also the desired quantities of antimicrobial and ant browning agents or any solute for improving sensory and nutritional quality. By controlling these above complex exchanges it is possible to conceive different combinations of water loss and solid gain, from a simple dewatering process (with substantial water removal and only marginal sugar pickup) to a candying or salting process (in which solute penetration in favored and water removal limited).

For porous foods, moist infusion can be also performed under vacuum. The internal liquid occluded in the open pores is exchanged for an external liquid phase (of controlled composition) due to pressure changes.

- When the food pieces are infused with the solutes and additives and then partially dried. The advantages obtained with this combination as compared to only drying are an increase in the stability of the pigments responsible for the color, an enhancement of the natural flavor, a better texture and a greater loading of the dryer.

Whatever the procedure used to reduce a_w, it is necessary to know the water activity-moisture content relationship in the food. Important contributions have been made in the field of a_w prediction over the past 50 years and comprehensive analysis of the procedures traditionally employed to calculate a_w have been performed by. In each case, the applicability of various theoretical and empirical equations was analyzed, presenting some descriptive examples.

There is no model with a simple mathematical structure capable of representing the sorption or aw lowering characteristics of foods or their components in the whole range of water activities, since the depression of a_w in foods is due to a combination of mechanisms each of which may be predominant in a given range of water activity. In high and intermediate moisture foods, a_w is mainly determined by the nature and concentration of soluble substances (i.e., sugars, NaCl, polios, amino acids, organic molecules, other salts) in the aqueous phase of food.

A number of equations, based on the thermodynamic properties of binary and multi components electrolyte and non-electrolyte solutions, have been studied theoretically and experimentally for calculating or predicting the a_w of these foods. Figure summarizes several of theoretical and empirical models suggested for the calculation of aw in semi-moist and moist food.

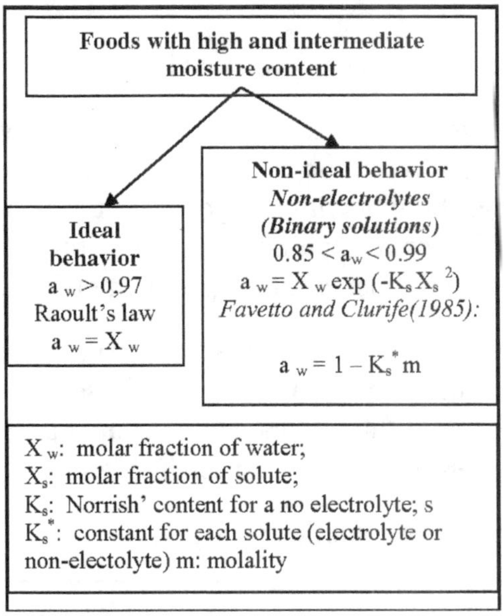

The applicability of various theoretical and empirical equations.

In low/moisture foods, adsorption of water rim surfaces is responsible for a_w reduction. Although the physical chemistry of surfaces has provided the food scientists with a large number of the critical equations, the relationship of water sorption -a_w cannot be predicted but must be experimentally determined due to many reasons. As food sorbs water, it can undergo changes of constitution, dimensions and other properties and sugars contained in the food may experience phase transformations.

The moisture sorption isotherm integrates the hygroscopic properties of numerous constituents whose sorption properties may change duet physical/chemical interactions induced by heating and other pre-treatments. Critical compilations of empirical and theoretical adsorption models for fitting experimental water sorption isotherms of food and food products.

The Brunauer, Emmett and Teller (BET) formula (application range $0.05 < a_w < 0.45$) is one of the most widely used models to characterize the monolayer water. The theory supposes, among other things, that the binding energy of the monolayer is the same for all the water molecules and on the other layers is equal to that of pure water.

Although the theoretical assumptions are incorrect for heterogeneous food surface interactions, for practical purposes this equation has been found very useful in determining the optimum moisture content (i.e., that corresponding to the monolayer water) for storage chemical stability of dehydrated foods.

The Guggenheim, Anderson and Boer (GAB) equation (applicability range $0 < a_w < 0.9$) now recognized as the most versatile sorption model and recommended as such by the European COST 90 Project, modifies the BET model to take into account the energies of interaction between the first and distant sorbet molecules at the individual sorption sites. Id also allows calculation of the monolayer water.

Recommended Equipment for Measuring a_w

Many methods and instruments are available for laboratory measurement of water activity in foods. Methods are based on the properties of solutions. Water activity can be estimated by measuring the following: Vapor pressure, Osmotic pressure, Freezing point depression of a liquid or solid, Boiling point elevation, Dew point and wet bulb depression, Suction potential, or by using the isopiestic method, electric hygrometers etc.

Fruits are a good example of foods stuffs that accept pH reduction without affecting the flavor significantly. Important development on IMF based on fruits and vegetables are reported elsewhere. The water reduction capacity of sugar and salts in their amorphous and anhydrous state at different aw in present in Table.

Table: Water activity reduction capacities of sugars.

Moisture content, gH_2O/100 g Solids	Sugars				
	Sucrose	Glu-cose	Fruc-tose	Lactose	Sorbitol (adsorption)
A_W	Anhydrous				
0.60	3.0	1.0	14.0	0.01	17.0
0.70	5.0	3.5	22.0	0.01	22.0

0.80	10.0	7.5	34.0	0.05	37.0
0.90	-	12.5	47.0	0.10	76.0
A_w	Amorphous				
0.60	14.0	1.0	18.0	4.5	25.0
0.70	20.0	3.5	30.0	4.7	35.0
0.80	35.0	8.0	44.0	4.7	55.0
0.90	65.0	22.0	70.0	-	110.0

To determine the desired a_w in syrup (aw equilibrium), the Ross equation is used:

$$a_{W\ equilibrium} = (a_w)\text{fruit x } (a_w)\text{ sugar}$$

Where a_w fruit is the water activity of the fruit and a_w sugar is the water activity of sugar, both calculated at the total molarities of the system. The product of the morality of sucrose in the fruit water and solution must equal the desired water activity in equilibrium. A_w values of the sugar are obtained using the Norrish equation:

$$a_{w\ sucrose} = X_1 \exp\left(-kX_2^2\right)$$

Where: k is a constant for sugars, X_1 and X_2 are the molar fractions of water and sugar, respectively. Some K values for common sugars and polyols are listed in table.

Phosphoric or citric acids are generally used to reduce the syrup's pH so that the final pH of the fruit-syrup system is in equilibrium in the desired range (3.0 to 4.1). Monitoring of a_w and pH in the fruit and syrup until constant values for these parameters are reached can determine the time to equilibrate the system. This may be from three to five days at constant room temperature depending on the size of fruit pieces.

Table: Norrish constant values for common sugars and polyols.

Sugars	k	Polyols	k
Sucrose	6.47 ±0.06	Glycerol	1.16 ±0.01
Maltose	4.54 ±0.02	Mannitol	0.91 ±0.27
Glucose	2.25 ± 0.04	Arabitol	1.41
Lactose	10.2	Propylene Glycol	4.04
Sorbitol	1.65 ±0.14		

Application of Norrish Equation

Example:

The water activity of a sucrose-water solution (2.44:1 w/w) can be estimated by means of the Norrish equation. The mole fractions are: $X_1 = 0.887$ and $X_2 = 0.1125$. The Norrish constant (k) for sucrose is 6.47. Substituting X_1 and X_2 into the Norris equation results in the estimated water activity of the sucrose-water solution.

To prepare the syrup or brine, a sufficient amount of sugar or salt is dissolved in water in order to reach the desired a_w. Concentrations of sculpture dioxide and potassium sorbate are prepared, reaching a final concentration of 100- 150ppm and 1000-1500ppm, respectively. In the case of fruit products, citric or phosphoric acid are used to lower the pH of the syrup so that the final pH at equilibrium is in the range 3.0-4.1. High moisture food products (HMFP) are very different from IMF products and need to be dehydrated. HMFP have a lower sugar concentration, 24-28% w/w compared to 20-40% w/w, and higher moisture content, 55-75% w/w compared to 20-40% w/w, which makes them similar to canned food products. HMFP can be consumed directly after processing or bulk stored for processing out of season.

Sample calculation for preparation of stable mango product, the process conditions and ingredients required to prepare 20kg of a stable mango product are: fruit pulp 16°Brix (16% soluble solids), acidity 0.5% (% citric acid). The fruit pulp is conditioned from 16°Brix (16% SS) to 40 °Brix (40% SS) by adding sucrose. Sucrose is added to the pulp in order to act as a water activity depressor. The water activity of the pulp ranges from 0.97 to 0.98. Calculation to obtain the amount of fruit pulp in the feed, sugar, citric acid, and free water in the final product:

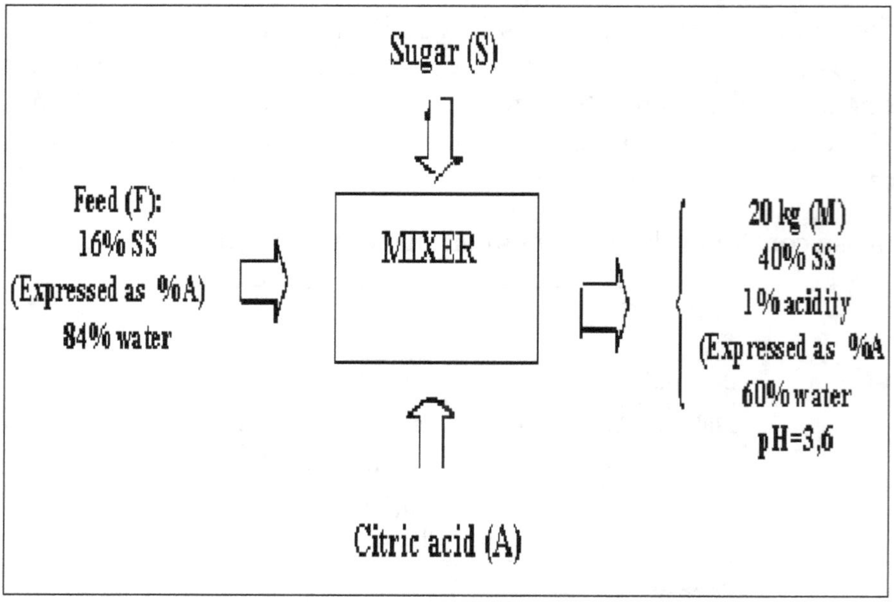

Modeling of stable mango product.

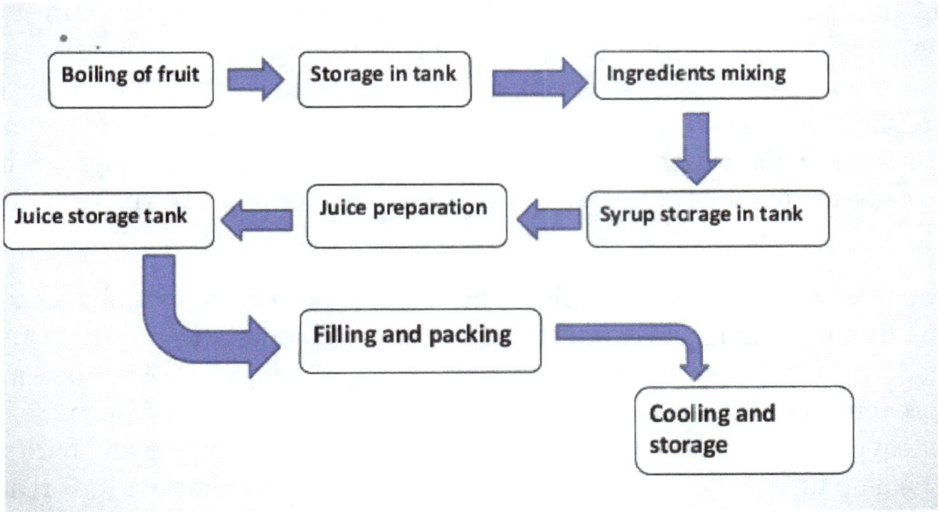

Flow process diagram for the food production.

The fruit prods were stored for at least 30 days at 35 °C, exhibiting door acceptability, microbial stability, and fresh-like appearance.

Quality and Food Security

Theoretical and experimental study of water activity concept allows us to evaluate the choice of method of preserving food. Further research needs to be targeted towards the stability in each macro–micro region and to explore more generic rules for stability.

Quality and food security depends on pH and water activity (a_w) in food environment. Foods with water activity are perishable. Water activity, *p*H, temperature, and other parameters, have a direct impact on the growth of microorganisms, thus a_w and *p*H are two the most important parameters. Free water that is available to molds, yeasts, and bacteria is responsible for their growth and even toxin production. Or it may participate in chemical/biochemical reactions (e.g. Maillard reactions), which might deteriorate: Texture, flavor, color, taste, nutritional value of a product, and its stability → shelf-life time.

Water activity (a_w) has its most useful application in predicting the growth of bacteria, yeasts and moulds. For a food to have a useful shelf life without relying on refrigerated storage, it is necessary to control either its acidity level (pH) or the level of water activity (a_w) or a suitable combination of the two. This can effectively increase the product's stability and make it possible to predict its shelf life under known ambient storage conditions.

Dehydration

Drying is a method of food preservation that works by removing water from the food, which prevents the growth of microorganisms and decay. It also slows down the action of enzymes, but does not inactivate them. Drying food using the sun and wind to prevent spoilage has been known since ancient times. Water is usually removed by evaporation, (air drying, sun drying, smoking or wind drying), but, in the case of freeze–drying, food is first frozen and then water is removed by sublimation. Because drying removes moisture, the food becomes smaller and lighter in weight. When the food is ready for use, the water is added back and the food returns to its original shape.

Dehydration of food is one of the oldest methods of preserving food for later use. It can either be an alternative to canning and freezing or a compliment to these methods. With modern food dehydrators, drying food is simple, safe and easy to learn. Dried food is great in traditional cooking recipes and can save you a lot of time in the kitchen during meal preparation time. Dried foods are also ideal for camping and backpacking as they take up little weight or space and do not require refrigeration. Kids really love fruit leathers too, which make a healthy nutritious snack food.

The process was developed commercially in France about 1795 to preserve sliced vegetables, using a hot-air blast. The earliest large-scale application was to starch products such as pasta, but after 1945 it was extended to milk, potatoes, soups, instant coffee, and prepared baby and pet foods. A major benefit to food manufacturers is reduction of weight and volume of the food products, thus lowering distribution cost.

The optimum temperature for drying food is 60 °C. If higher temperatures are used, the food will cook instead of drying. When the food cooks on the outside and the moisture cannot escape, "case hardening" can occur. The food will eventually mold. Thus, the drying process should never be hurried by raising the drying temperature.

Low humidity aids the drying process. Food contains a lot of water. To dry food, the water must move from the food to the surrounding air. If the surrounding air is humid, then drying will be slowed down.

Increasing the air current speeds up drying by moving the surrounding moist air away from the food. To speed the drying time, increase the air flow.

Many different foods are prepared by drying, including Parma ham, bresaola, beef jerky, and even fruits that normally have a high water content, such as prunes, raisins, figs, and dates.

Dried and salted reindeer meat is a traditional Lappish food. First the meat is soused. It is kept in saltwater for a couple of days to guarantee the conservation of the meat. Then the meat is dried in the sun in spring when the air temperature is below zero. The dried meat can be further processed to make soup.

Dehydrated food can be fun (as in "fruit leather"), chic (sundried tomatoes) and handy (herbs). It is also easy to make and store.

Principles of Food Dehydration

Moisture in Foods

Basically, water may be held in a food product in two distinct ways. Some water may be held in the interstitial spaces and within the pores of the material by purely physical force related to surface tension. This type of water is called "unbound water." It exerts the same vapor pressure and possesses the same latent heat of vaporization as does pure water at the same temperature. The amount of unbound water in a food product is more closely related to the physical structures of the product than to the chemical structures of the product.

Another portion of water may be held on the internal and external surfaces of the solid material by interactions between the water molecules and those of the solid material to form a mono-layer of water molecules. Some water may be held by interactions between the water molecules to form a multi-layer of water molecules. This type of water is called "bound water," which exerts a vapor pressure less than that of pure water at the same temperature. Also, the heat of vaporization for bound water is greater than that of pure water at the same temperature. The amount of bound water held in a food product is closely associated with the chemical structures of the product. The inter-molecular forces known as "van der Waals" forces are responsible for holding water on sorption sites on the adsorbent. In general, polar molecules such as H_2O, NH_3 and alcohol, or molecules possessing the following polar groups $-NH_2$, -NH ·, -OH, -COOH, $-CONH_2$, etc., are considered to be sorptive sites on adsorbent.

Free moisture refers to the moisture contained by a product in excess of the equilibrium moisture content of a product. Only the free moisture in a product can be removed during a given dehydration process. The free moisture of a product depends upon the type of product, and temperature and water vapor concentration in the air.

The bound water in a food product is removed with a certain degree of difficulty during a dehydration process. A plot of the heat of desorption with respect to the moisture content of a product would describe approximate ranges of both bound and unbound water in a product, based on the value of the heat of desorption. The heat of desorption also indicates the binding energy or the inter-molecular forces between water molecules and the surface of the adsorbent, and between adsorbed water molecules and molecules of water vapor. The energy required to evaporate the unbound water (above 23% moisture, d.b.) in corn starch is almost the same as the heat of vaporization of pure water. However, the energy required to evaporate the bound water is much greater than the heat of vaporization of pure water, especially in the lower moisture range. Sometimes, conditions that will remove all the bound water may cause some chemical decomposition, and a resultant weight loss of dry matter.

Equilibrium Moisture Relationship

The nature of the sorption relation between food products and water vapor is of great practical importance in the food industries, particularly in dehydration and quality preservation. The moisture in a food product exerts a partial pressure (P). The equilibrium moisture content (M_e) of a substance is the moisture content at which moisture in a product is in equilibrium with that in its surrounding (partial pressure of water in a product that of water vapor in air). The ratio of the equilibrium partial pressure to the vapor pressure of pure water (P_0) at a given temperature is known as the equilibrium relative humidity (P/P_o J or water activity, a_w.

The most convenient method to study properties of water involves preparation of "sorption isotherms," that is, curves relating the equilibriums moisture content to the relative humidity or water activity at a constant temperature. If the partial pressure of water in a product at a given moisture content is higher than the partial pressure of water vapor in the surrounding air, the product will lose water until equilibrium is achieved (desorption). If the partial pressure of water in a product is lower than the equilibrium partial pressure of water vapor in the surrounding air, the product will gain moisture from the surrounding air (adsorption).

Many products exhibit different equilibrium moisture characteristics depending upon whether the equilibirum is reached by desorption or adsorption of the moisture. In a dehydration operation, the equilibrium moisture by desorption is of particular interest. In considering a possible regain of moisture by dried products in storage, the equilibrium moisture by adsorption is of interest. In general, the equilibrium moisture by desorption is higher than that by adsorption.

From a theory suggested by, a well-known relationship between relative humidity and amount of adsorbate has been determined. The resulting expression, which is known as the BET equation, can describe the isotherms for many products:

$$\frac{P}{M(P_0 - P)} = \frac{1}{M_m C} + \frac{C-1}{M_m C} \bullet \frac{P}{P_0}$$

Where P =partial pressure of water vapour:

P_o = vapor pressure of pure water at a given temperature.

M = moisture content, lb of H_zO/lb of dry solid.

M_m = moisture content corresponding to a monolayer of adsorbed moisture.

C = energy constant, $\frac{E_1\text{-}L}{RT}$.

E_1 = heat of adsorption of molecules in mono-layer.

L = latent heat of condensation.

R = universal gas constant.

T =absolute temperature.

A plot of $P/M(P_o-P)$ vs. P/P_o results in a straight line of slope, $C-1/M_mC$ and intercept, $1/M_mC$ if a product follows the BET theory. The BET equation usually can be expected to describe equilibrium moisture content data up to 40% equilibrium relative humidity. From the BET equation, the moisture content corresponding to a mono-layer can also be evaluated.

The isotherm equations developed by Henderson, and Chung and Pfost describe the relationship between equilibrium moisture content and relative humidity or water activity for many agricultural and food products in a wide range of relative humidity or water activity.

The isotherm equation developed by Henderson is given as:

$$1-\frac{P}{P_o}=\exp\left(-cTM_e^n\right),$$

where, M_e = equilibrium moisture content, dry basis %.

P/P_o =equilibrium relative humidity or water activity.

T =absolute temperature.

c and n = constants.

The equation developed by Chung and Pfost (4) is:

$$\ln\frac{P}{p_o}=\frac{A}{RT}e^{-BM_e}$$

where, A and B = constants.

R = universal gas constant.

For preservation of product quality, water activity rather than moisture content is of importance because water activity, a_w, determines the lower limit of available water for microbial growth. Most bacteria do not grow below $a_w = 0.8$. Some fungi have been reported to grow at $a_w = 0.65$, but the range of 0.7 - 0.75 is generally considered their lower limit.

Dehydration Rate Mechanism

Dehydration is a process of simultaneous heat and moisture transfer. The heat is required to evaporate the moisture, which is removed from the product surface by the external dehydration medium, usually air. The dehydration process can be divided into two major rate periods: (a) the constant-rate period and (b) the falling-rate period.

Constant-rate Period

If a product is initially very wet, the surface will be covered with a thin film of water, which is entirely unbound water. When it is exposed to relatively dry air, evaporation will take place from the surface. In this initial dehydration period, the rate of evaporation or dehydration remains constant until the average moisture content of the product reaches a value M_e, the critical moisture content. The constant-rate period may be missing entirely, depending upon the type of product and the initial moisture content of product.

In the beginning, the product and the water surface are usually colder than the ultimate surface temperature and the evaporation rate will increase while the surface temperature rises to its ultimate value. Alternatively, the ultimate surface temperature may be lower than the initial value. The initial period is usually so short that it is ordinarily ignored in subsequent analysis of the dehydration process.

In the constant-rate period, the dehydration rate can be described in terms of mass-transfer coefficient K_m and the difference in humidity of the air at the surface and in the main stream, or balancing the heat requirement for evaporation and the rate at which heat reaches the surface.

$$\frac{dM}{d\Theta} = K_m A(H_S - H_a) = \frac{h_c A(t_a - t_S)}{h_{fg}},$$

where, $\frac{dM}{d\Theta}$ = dehydration rate, lb of water evaporated/h.

M = moisture content of a product, lb of water/lb of dry solids.

Θ = time, h.

A = dehydration surface area, ft^2.

t_S = temperature of surface, °F.

t_a = temperature of air, °F.

H_S =humidity of air at t_S, lb of water vapor/lb of dry air.

H_a =humidity of air at ta, lb of water vapor/lb of dry air.

K_m = mass-transfer coefficient at the water-air interface, lb of water evaporated/ $ft^2 - h - \Delta H$.

h_e = heat transfer coefficient at the water-air interface, $BTU/h\text{-}ft^2 - {}^\circ F$.

h_{fg} = latent heat of vaporization at t_S, Btu/lb of water.

Falling-rate Period

When the average moisture content of the product reaches the critical moisture content, the surface film of water is so reduced by evaporation that further dehydration causes dry spots to appear on the surface and these occupy increasingly larger proportions of the exposed surface as dehydration proceeds. This gives rise to the first part of the falling-rate period, the period of unsaturated surface dehydration. Ultimately the original surface film of water is entirely evaporated. On further dehydration, the rate at which moisture may move through the product is the controlling step, as a result of moisture gradients existing between the internal parts and the surface.

As the moisture content generally is lowered by the dehydration, the rate of internal moisture movement decreases. The rate of dehydration falls even more rapidly than before. At a point the moisture content of the product reaches the equilibrium moisture content Me for a given dehydration conditions and dehydration stops.

For the unsaturated surface dehydration in the falling-rate period, the dehydration rate can be described by the equation:

$$dM/d\Theta = aM + b$$

Where a and b are constants.

For the second part of falling-rate period (internal moisture movement to the surface), the dehydration rate can be described by the Fick's second law of diffusion:

$$\frac{dM}{d\Theta} = D_V\left[\frac{\delta^2 M}{\delta r^2} + \frac{c\,\delta M}{r\,\delta r}\right]$$

Where c = constant (c = 0 for planar symmetry, c = 1 for cylindrical body, c = 2 for spherical body).

r = product coordinate.

D_v = diffusion coefficient, $\frac{ft^2}{h}$.

The initial and boundary conditions that are commonly used in describing the falling-rate period during a food dehydration process are:

$$M(r,o) = M_O, \text{at } \Theta = 0.$$

$$M(r_O,\Theta) = M_e, \text{at } r = r_O \text{ (at the surface)}.$$

$$M(0,\Theta) = \text{finite, at } r = 0 \text{ (at the center)}.$$

The analytical solutions of the Fick's second law of diffusion equation with the above initial and boundary conditions (*I*) are:

For an inftnite plane,

$$\frac{\overline{M}-M_e}{M_o-M_e}=\frac{8}{1^2}\sum_{n=0}^{\infty}\frac{1}{(2n+1)^{2\exp}}\left[-\frac{(2n+1)^2\delta^2}{4}x^2\right]$$

For a sphere,

$$\frac{\overline{M}-M_e}{M_o-M_e}=\frac{6}{\delta^2}\sum_{n=1}^{\infty}\frac{1}{n^2}\exp\left[-\frac{n^2\delta^2}{9}x^2\right]$$

For an inftnite cylinder,

$$\frac{\overline{M}-M_e}{M_o-M_e}=\sum_{n=1}^{\infty}\frac{4}{\tilde{a}_n^2}\exp\left[-\frac{\tilde{a}_n^2}{4}x^2\right]$$

Where, $X=\frac{A}{V}(D_V\,\Theta)^{1/2}$

A = surface area of a product, ft².

V = volume of a product, ft³.

$\overline{M}$ =average moisture content of a product ate = Θ , lb of water/lb of dry solid.

M_o = initial moisture content, lb of water/lb of dry solids.

M_e =equilibrium moisture content, lb of water/lb of dry solid.

n = integer.

$\tilde{a}_n$ = roots of the Bessel function of zero order.

A simplified version of the above three equations is frequently used to predict the dehydration rate. Instead of an infinite number of terms, only the first term of the equations is employed to calculate the dehydration rate. The resulting expression is:

$$\frac{\overline{M}-M_e}{M_o-M_e}=Ae^{-k\Theta}$$

Where, A = constant.

K =dehydration constant, h^{-1}.

Because of complexity and diversity of food products, many types of dehydrators were developed to meet dehydration characteristics of specific food products. In selecting a dehydrator, one should carefully consider the following factors: (a) capacity (b) thermal efficiency and (c) product quality. Whatever the dehydration method employed, an understanding of basic theories and principles involved in the dehydration process together with operational procedures of a given dehydrator is important for solving dehydration problems encountered in practice.

Concentration of Moist Foods

Foods with substantial acidity, when concentrated to 65 percent or more soluble solids, may be preserved by mild heat treatments. High acid content is not a requirement for preserving foods concentrated to over 70 percent solids.

Fruit jelly and preserve manufacture, an important fruit by-product industry, is based on the high-solids–high-acid principle, with its moderate heat-treatment requirements. Fruits that possess excellent qualities but are visually unattractive may be preserved and utilized in the form of concentrates, which have a pleasing taste and substantial nutritive value.

Jellies and other fruit preserves are prepared from fruit by adding sugar and concentrating by evaporation to a point where microbial spoilage cannot occur. The prepared product can be stored without hermetic sealing, although such protection is useful to control mold growth, moisture loss, and oxidation. In modern practice, vacuum sealing has replaced the use of a paraffin cover.

The jelly-forming characteristics of fruits and their extracts are due to pectin, a substance present in varying amounts in all fruits. The essential ingredients in a fruit gel are pectin, acid, sugar, and water. Flavouring and colouring agents may be added, and additional pectin and acid may be added to overcome any deficiencies in the fruit itself.

Candied and glacéed fruits are made by slow impregnation of the fruit with syrup until the concentration of sugar in the tissue is sufficiently high to prevent growth of spoilage microorganisms. The candying process is conducted by treating fruits with syrups of progressively increasing sugar concentrations, so that the fruit does not soften into jam or become tough and leathery. After sugar impregnation the fruit is washed and dried. The resulting candied fruit may be packaged and marketed in this condition or may be dipped into syrup, becoming coated with a thin glazing of sugar (glacéed) and again dried.

Methods of Concentration

Solar Evaporation

Solar evaporation is the simplest method of evaporating water with solar energy. This process was used in earlier times to obtain salt from sea water and still it is practiced. However, the process is very slow and is suitable only for concentrating salt solutions.

Open Kettles

Only some foods can be satisfactorily concentrated in open kettle that is heated by steam e.g. in case of jellies and jams and for certain types of soups. However, high temperatures and long concentration times damage most foods. In addition, thickening and burning of product to the kettle wall gradually lowers the efficiency of heat transfer and slows the concentration process. This method is apt for caramelized colour and typical flavour development in foods high in sugar.

Concentration by Flash Evaporation

Concentration process is markedly speeded when sub sized food material is brought in direct contact with heating medium. This is done in flash evaporators. Clean steam superheated at about 150 °C is injected into food and then is pumped into an evaporation tube where boiling occurs. The boiling mixture then enters a separator vessel and the concentrated food is drawn off at the bottom and the steam plus water vapour from the food is evacuated through a separate outlet. Foods lose volatile flavour constituent because of high temperature.

Concentration by Thin Film Evaporation

In thin film evaporators, food is pumped into a vertical cylinder which has a rotating element that spreads the food into a thin layer on a cylinder wall. The cylinder wall of double jacket construction usually is heated by steam. Water is quickly evaporated from the thin food layer and the concentrated food is simultaneously wiped from the cylinder wall. The concentrated food and water vapour are continuously discharged to an external separator from which product is removed at the bottom and water vapour passes to a condenser. Product temperature may reach 85 °C or higher but since residence time of the concentrating food in the heated cylinder may be less than a minute, heat damage is minimal.

Concentration by Vacuum Evaporation

This method is suitable for heat sensitive foods as this method involves low temperature. Evaporation under vacuum can be done by operating thin film evaporators under vacuum by connecting a vacuum pump or steam ejector to the condenser. Several vacuum vessels can be attached in series so that the food product moves from one vacuum chamber to the next and thereby becomes more and more concentrated at each step. The consecutive vessels are maintained at progressively higher degrees of vacuum and hot water vapour arising from first step is used to heat the second vessel and so on. In this way heat energy is efficiently used.

Freeze Concentration

Initially formed ice crystals during freezing process are removed with the help of centrifugal force resulting in a concentrated unfrozen food which passes through a fine mesh screen. This process is repeated many times to reach final concentration of food.

Ultrafiltration and Reverse Osmosis

These are the two methods of concentrating foods employing pressure driven membrane separation process. In ultrafiltration large solute particles are selectively removed whereas in reverse osmosis smaller solutes are separated out.

Intermediate Moisture Foods (IMF)

Intermediate moisture foods are those in which the moisture content is reduced to a level low enough to prevent spoilage microorganisms from growing but moist enough for the food to have improved palatability characteristics. Intermediate moisture foods or semi-moist foods contain 20-50 percent. In addition, they contain high concentration of dissolved solutes. These foods do not require refrigeration during storage and can be eaten without rehydration. Honey, jam, jelly, cakes, dates and osmo-dried food products are the examples of intermediate moisture foods.

Fermentation

Fermentation is an essential metabolic phenomenon that basically takes place in absence of oxygen (O_2). In the process of fermentation, sugar is consumed in the absence of oxygen. The products formed due to fermentation are organic acids, gases, or alcohol. Fermentation occurs commonly in yeast and bacteria and also in oxygen-starved muscle cells, as in the case of lactic acid fermentation. In terms of microbiologists, fermentation is a primary means of producing ATP by the degradation of organic nutrients anaerobically, in presence of suitable microorganisms. Zymology is the science of fermentation. Fermentation processes are believed to have been developed in order to preserve fruits and vegetables for times of scarcity by preserving the food by organic acid and alcohols, impart desirable flavour, texture to foods, reduce toxicity and decrease cooking time. When the fermentation term is used in case of fruits and vegetables, it is known as pickling. Fermented fruits and vegetables have an important role in feeding the world's population. It is well documented that fermented drinks were consumed in Babylon 5000 years ago and that bread was consumed in Egypt around the 1500 BC . Fermentation is employed in the production of foods through the application of microorganisms or their enzymes . Fermented foods are food substrates that are invaded or overgrown by edible microorganisms whose enzymes, particularly amylases, proteases; lipases hydrolyze the polysaccharides, proteins and lipids to nontoxic products with flavors, aromas and textures pleasant and attractive to the human consumer . Humans have used fermentation to produce drinks and beverages since the Neolithic age. For example, fermentation is used for preservation in a process that produces lactic acid as found in such sour foods as pickled cucumbers, kimchi and yogurt, as well as for producing alcoholic beverages such as wine and beer. Fermentation occurs within the gastrointestinal tracts of all ruminant animals, including humans.

Biochemistry of Fermentation

Fermentation is essentially takes place in anaerobic conditions when there is no oxidative phosphorylation to maintain the production of ATP (Adenosine triphosphate) by glycolysis. During fermentation pyruvate is metabolized to diverse compounds.

3 types of fermentation usually take place:

- Homo-lactic fermentation is the production of lactic acid from pyruvate.
- Hetero-lactic fermentation is the production of lactic acid as well as other acids and alcohols.
- Alcoholic fermentation is the conversion of pyruvate into ethanol and carbon dioxide. Some typical examples of fermented products are ethanol, lactic acid, and hydrogen. Several other exotic compounds can be produced by fermentation that includes butyric acid and acetone.

The final step of fermentation i,e. conversion of pyruvate to fermentation end-products does not produce any energy. This is critical for an anaerobic cell since it rejuvenates nicotinamide adenine dinucleotide (NAD), which is required for the glycolysis process. This is having imperative role for normal cellular function, as glycolysis is the only source of ATP in anaerobic conditions.

Fermentation turns NADH and pyruvate produced in glycolysis into NAD^+ and an organic product. In the presence of O_2, NADH and pyruvate are used to generate ATP in the process of respiration. This process is known as oxidative phosphorylation. It generates much more ATP as compared to glycolysis alone. The magical process of fermentation is rarely utilized when oxygen is available. The process is necessarily anaerobic which means that it only takes place in absence of oxygen (O_2). Obligate anaerobes are those anaerobes which cannot tolerate oxygen at even physiological range of concentration.

The first step is the Embden – Meyerof - Parnas glycolysis which is same in most of the fermentation pathways:

$$C_6H_{12}O_6 + 2\,NAD^+ + 2\,ADP + 2\,P_i \rightarrow 2\,CH_3COCO_2^- + 2\,NADH + 2\,ATP + 2\,H_2O + 2\,H^+$$

Pyruvate is $CH_3COCO_2^-$. Pi is inorganic phosphate. Two ADP molecules and two Pi are converted to two ATP and two water molecules via substrate-level phosphorylation. Two molecules of NAD^+ are also reduced to NADH.

In oxidative phosphorylation, the energy for ATP formation is derived from an electrochemical proton gradient generated across the inner mitochondrial membrane (or, in the case of bacteria, the plasma membrane) via an electron transport chain. Glycolysis has substrate-level phosphorylation (ATP generated directly at the point of reaction).

The reaction differs according to the sugar being used in the process of fermentation, as well as the particular organism performing it. Below, the sugar is glucose ($C_6H_{12}O_6$), the most common sugar and the process is the alcoholic fermentation.

Chemical Equation:

$$C_6H_{12}O_6 \rightarrow 2\,C_2H_5OH + 2\,CO_2 + 2\,ATP \text{ (Released energy : 118 kJ mol}^{-1}\text{)}$$

Word Equation:

Sugar (glucose) → Alcohol (ethanol) + Carbon Dioxide (CO_2) + Energy (as ATP).

Types of Fermentation

Ethanol Fermentation

Ethanol fermentation is the process in which glucose is converted to ethanol (C_2H_5OH) and carbondioxide (CO_2). Ethanol fermentation is also known as alcoholic fermentation that converts sugars such as glucose, fructose, and sucrose into energy (ATP), ethanol and carbon dioxide as by-products. One glucose molecule is converted into two ethanol molecules and two carbon dioxide molecules. yeasts perform this conversion in the absence of oxygen, alcoholic fermentation is considered an anaerobic process.

$$C_6H_{12}O_6 \rightarrow \text{Pyruvate} \rightarrow 2\,C_2H_5OH + 2\,CO_2$$

Ethanol fermentation/yeast fermentation have wide practical applications in the production of alcoholic beverages, ethanol and bread. This process is mainly responsible for the existence of any sort of alcohol. Man have harnessed the potential of controlled bio-fermentation to achieve huge potential wine industry, bread industry and cheese industry.

Basically the reaction takes place in 3 steps:

Step 1: Enzyme invertase cleaves the glycosidic linkage in disaccharide sucrose between the glucose and fructose molecules in.

$$C_{12}H_{22}O_{11} + H_2O + \text{invertase} \rightarrow 2\,C_6H_{12}O_6$$

Step 2: Glucose molecule is broken down into two pyruvate molecules (glycolysis).

$$C_6H_{12}O_6 + 2\,ADP + 2\,Pi + 2\,NAD^+ \rightarrow 2\,CH_3COCOO^- +$$
$$2\,ATP + 2\,NADH + 2\,H_2O + 2\,H^+$$

Step 3: Pyruvate is converted to ethanol and CO_2 (by two step reaction).

$CH_3COCOO^- + H^+ \rightarrow CH_3CHO + NADH + H^+ \rightarrow C_2H_5OH + NAD^+$ (in presence of pyruvate decarboxylase and alcohol dehydrogenase).

Lactic Acid Bacteria Fermentation

Out of the various approaches to fermentation, lactic acid fermentation is strictly controlled by salt concentration. Lactic acid bacteria fermentation is done by using natural microflora or lactic acid bacterial (LAB) cultures which is employed throughout the world, in conjunction with chemical preservation, using salt and acid to preserve various foods such as milk, cereals, meat, and fruits and vegetables. Lactic acid (LA) fermentation of vegetables and fruits is a common practice to maintain and improve the nutritional and sensory features of food commodities Lactic acid fermentation retains all the natural plant ingredients while improving the quality, taste and aroma. LA fermentation enhances the organoleptic and nutritional quality of the fermented fruits and vegetables and retains the nutrients and coloured pigments. Fermentation plays at least five roles in food processing:

- Enrichment of the human dietary through development of a wide diversity of flavors, aromas and textures in food.
- Preservation of substantial amounts of food through lactic acid, alcoholic, acetic acid, alkaline fermentations and high salt fermentations.
- Enrichment of food substrates biologically with vitamins, protein, essential amino acids and essential fatty acids.
- Detoxification during food fermentation processing.
- Decrease in cooking times and fuel requirements.

Lactic acid fermentation is used in many areas of the world to produce foods that cannot be produced through other methods. The most commercially important genus of lactic acidfermenting bacteria is Lactobacillus, though other bacteria and even yeast are sometimes used. Two of the most common applications of lactic acid fermentation are in the production of yogurt and sauerkraut. Lactic acid is naturally produced in fermented products from the sugar present in the fruit sample. Lactic acid bacteria is a group of gram positive, non-spore forming, cocci or rod which produce lactic acid as a major end product. The basic mechanism of the preservation of foods is the production of acid, chiefly by LAB, which lowers the pH to a level at which most of the spoilage-causing microorganisms cannot grow, and, thus, the food is preserved. Food substrates overgrown with desirable, edible microorganisms become resistant to invasion by spoilage, toxic or food poisoning microorganisms. Other, less desirable or pathogenic organisms find it difficult to compete. The consumption of LA fermented fruits and vegetables helps to enhance human nutrition in several ways such as the attainment of balanced nutrition, providing vitamins, minerals, and carbohydrates, and preventing several diseases such as diarrhoea and cirrhosis of liver because of probiotic properties.

Probiotic is a relatively new word meaning "for life" and it is generally used to name the bacteria associated with beneficial effects for humans. Probiotics are defined as live microbial feed such as Lactobacillus plantarum, L. casei, L. acidophilus, and Streptococcus lactis which are supplemented by food that beneficially affect the host by improving its intestinal balance. Several studies have shown that supplementation of probiotics to food provides several health benefits such as reduction of serum cholesterol, improved gastrointestinal function, enhanced immune system, and lower risk of colon cancer. A number of studies have found probiotics consumption to be useful in the treatment of diarrhea, lactose intolerance, colon cancer, cholesterol, blood pressure, immune function and infections, mineral absorption, irritable bowel syndrome and colitis. Important probiotic bacteria can be listed as Lactobacillus plantarum, Lactobacillus acidophilus, Lactobacillus rhamnosus, Bifidobacterium bifidum, Bifidobacterium longum, Pediococcus acidilactici and Saccharomyces boulardii. The genus Lactobacillus is a heterogeneous group of LAB with important application in food and feed fermentation. Lactobacilli are used as probiotics inoculants and as starters in fermented food. The genus Lactobacillus is Gram-positive organisms which produce lactic acid by fermentation which belongs to the large group of LAB. Other genera such as Lactococcus, Enterococcus, Oenococcus, Pediococcus, Streptococcus, Leuconostoc, and Lactobacillus are also considered in LAB group due to lactic acid production ability.

The LAB tolerate high salt concentrations, which give them an advantage over other less salt-tolerant species and allows the LAB to produce acid that inhibits the growth of undesirable microorganisms. Leuconostoc sp. is also known for its high salt tolerance, and, for this reason, initiates the majority of lactic acid fermentations. The addition of salt to the pickles restricts the growth of gram-negative bacteria and enhances the growth of LAB and Leuconostoc sp. LAB are one of the important microorganisms in food fermentation, and have been shown by serological techniques and 16'S ribosomal RNA cataloging to be phylogenetically related and to share a number of common features. Lactic acid fermentation increases shelf life of fruits and vegetables and also enhances several beneficial properties, including nutritive value and flavour, and reduces toxicity. LAB are recognized for their fermentative ability and thus enhancing food safety, improving organoleptic attributes, enriching nutrients and increasing health benefits. Fermented fruits and vegetables can be used as a potential source of probiotics as they harbor several lactic acid bacteria such as Leuconostoc mesenteroides, Lactobacillus brevis, Lactobacillus plantarum, Pediocccus cerevisiae, Streptococcus thermophilus, Streptococcus lactis, Lactobacillus bulgaricus, Lactobacillus acidophilus, Lactobacillus citrovorum, Bifidobacterium bifidus and L. pallax while Staphylococcus aureus, Saccharomyces cerevisiae, and A. niger may be present when a pickle is spoiled. Spoilage of pickles can also be due to microbial contamination or oxidation rancidity of the oil used.

Some products made by harnessing the principle of Lactic Acid Bacteria fermentation includes:

- Pickle – It's a product prepared by lactic acid bacteria (LAB) fermentation of sugar present in pieces of fruits and vegetables. The prepared product is rich in Lactic acid and only the beneficial bacteria that can tolerate lactic acid pH survive.
- Sauerkraut – It is basically a finely cut cabbage that has been fermented by various lactic acid bacteria. Sauerkraut usually has a long shelf life and a distinctive sour flavor.
- Yogurt – Yogurt is basically a fermented product prepared from milk. The main method of producing yogurt is through the lactic acid fermentation of milk with harmless bacteria. The primary bacteria used are typically Lactobacillus bulgaricus and Streptococcus thermophiles.
- Kimchi – It is basically a Korean dish. It is a staple food in Korean cuisine, is a traditional side dish made from salted and fermented vegetables, most commonly napa cabbage and Korean radishes, with a variety of seasonings including chili powder, scallions, garlic, ginger, and jeotgal.

Acetic Acid Bacteria Fermentation

AAB are a group of gram-negative bacteria which oxidize sugars or ethanol and produce acetic acid during fermentation. The acetic acid bacteria consist of 10 genera in the family Acetobacteraceae. Several species of acetic acid bacteria are used in industry for production of certain foods and chemicals. Vinegar is formed when acetic acid bacteria is added to alcoholic beverages. In this process, oxidative fermentation takes place that creates vinegar as a by-product. This process is somewhat aerobic. Weakly fermented liquors very often become sour on exposure to the air. This is owing to the conversion of the alcohol in acetic acid. Acetic acid is produced by fermenting various substrates (starchy solution, sugar solutions or alcoholic foodstuffs such as wine or cider) in presence of Acetobacter bacteria. Acetobacter aceti is usually used to produce vinegar (14 per cent acetic acid).

Starchy Solution/Sugar solutions/Alcohol + $O_2 \rightarrow CH_3COOH$

Different Fermented Products

Fermented products are of immense importance to mankind since time immemorial. Fermented products are not only a source of probiotics but also have good palatable qualities. Various products produced by the process of fermentation are actually waste products produced during the reduction of pyruvate to regenerate NAD^+ in the absence of oxygen. Bacteria generally produce acids by fermentation process. Vinegar (5% acetic acid) is the direct result of bacterial metabolism. In the mechanism of vinegar production, Bacteria convert the alcohol to acetic acid. In milk, the acid coagulates the casein to produce curds. In pickling, due to the presence of salt only the lactic acid bacteria survive. The LAB actually converts the sugar present in the pickling material and produce lactic acid. The acid thus produced, preserves the food from pathogenic and putrefactive bacteria.

When yeast fermentation takes place it breaks down the glucose ($C_6H_{12}O_6$) into exactly two molecules of ethanol ($C_2H_6O_5$) and two molecules of carbon dioxide (CO_2). Ethanol fermentation breaks the pyruvate down into ethanol and carbon dioxide. This is a mandatory process in bread-making, brewing, and wine-making. When the ferment has a high concentration of pectin, minute quantities of methanol can be produced. Usually only one of the products is desired; in bread the alcohol is baked out, and in alcohol production the carbon dioxide is released into the atmosphere. Lactic acid fermentation breaks down the pyruvate into lactic acid in presence of Lactic Acid bacteria. It takes place in the muscles of animals when they need energy faster than the blood can supply oxygen. It generally takes place during vigorous exercise. It also takes place in bacteria and fungi. It is this type of bacteria that convert lactose into lactic acid in yogurt, giving it its sour taste.

Pickling

Pickling is the process by which fresh fruits and vegetables are preserved and with the addition of salt, chilly and spices, a tasty preparation known as "Pickles" is made.

Pickles are also good appetizers and digestive agents. There are several varieties of pickles and they are consumed throughout the year by people from all walks of life. Unimaginable quantities of pickles are consumed round the year. On an average, each family consumes about 2 kgs of pickles every year. Pickles, crisp and spicy, stimulate the sense of taste and enhance the flavour of bland foods. Pickles and relishes contain small amounts of nutrients, depending on ingredients used in making them. Most pickle products are low in calories, except for the sweet varieties.

Pickling is the process of preserving foods in brine or vinegar or a combination of the two. Brine is made by combining salt with water in proportions to make either a weak, medium or strong solution. In some instances, salt is added directly to the food in the dry form and the brine is formed as juices are drawn out of the food.

Vinegar, an acid, acts as a preservative and contributes flavour different from the flavour produced by lactic acid fermentation that occurs during the brining process. The method using both salt and vinegar calls for a short brining period before the vinegar is added. Kinds of pickles and relishes are varied and numerous. Processing methods for each should be selected in keeping with the food to be processed and the desired product.

Pickling, also known as brining or corning is the process of preserving food by anaerobic fermentation in brine (a solution of salt in water) to produce lactic acid, or marinating and storing it in an acid solution, usually vinegar (acetic acid). The resulting food is called a pickle. This procedure gives the food a salty or sour taste. Pickling is preserving a food with acid and salt. The key to safe pickling is making sure that the acid is high enough to kill any microorganism that can lead to spoilage and illness.

Chemical Preservation

Salt and Sugar

Salt and sugar have long been used as effective means of extending shelf life of various products as these solutes bind water, leaving less water available for the growth of microorganisms. Essentially the water activity (a_w) of the product is reduced, and since most microorganisms require a high water activity, they areunable to survive.

Salt and sugar in concentrated solutions have high osmotic pressure. When these are sufficient to draw water from microbial cells or prevent normal diffusion of water into these cells, a preservative condition exists.

The critical concentration of sugar in water to prevent microbial growth will vary depending upon the type of micro-organisms and the presence of other food constituents, but usually 70% sucrose in solution will stop growth of all micro-organisms in foods. Less than this concentration may be effective but for short period of time unless the foods contain acid or they are refrigerated.

Salt becomes a preservative when its concentration is increased and levels of about 18% to 25% in solution generally will prevent all growth of micro-organisms in foods. Except in the case of certain briny condiments, however, this level is rarely tolerated in foods.

Acids

Acids such as citric acid, acetic acid (vinegar) and ascorbic acid are also known to confer protection against product deterioration. In these cases, the pH of the product is shifted to being low, that is, more acidic, where very few molds, yeast and bacteria are able to grow and multiply.

Acetic acid is a general preservative inhibiting many species of bacteria, yeasts and to a lesser extent moulds. It is also a product of the lactic-acid fermentation, and its preservative action even at identical pH levels is greater than that of lactic acid. The main application of vinegar (acetic acid) includes products such as pickles, sauces and ketchup.

Other Acidulants

- Malic and tartaric (tartric) acids is used in some countries mainly to acidify and preserve fruit sugar preserves, jams, jellies, etc.
- Citric acid is the main acid found naturally in citrus fruits; it is widely used (in carbonated beverages) and as an acidifying agent of foods because of its unique flavour properties. It has an unlimited acceptable daily intake and is highly soluble in water. It is a less effective antimicrobial agent than other acids.
- Ascorbic acid or vitamin C, its isomer isoascorbic or erythorbic acid and their salts are highly soluble in water and safe to use in foods.

Preservatives

Chemical food preservatives are those substances which are added in very low quantities (up to 0.2%) and which do not alter the organoleptic and physico-chemical properties of the foods at or only very little.They are used to improve the colour and keeping qualities of the final product for some fruits and vegetables.

Food additives such as benzoate and sorbate are quite commonly used in the fruit drink industry to protect against microbial spoilage, while nitrites are used in meat processing. These chemicals work best at acidic pH ranges and when the products are pasteurized. A combination of heat and chemicals where applicable is usually more effective than either one on its own.

Treatment with preservatives takes place after blanching or, when blanching is not needed, after slicing. The composition and strength of the preservative solution vary for different fruit and vegetables. As a general rule, preservatives are not used for treating onions, garlic, leeks, chilies and herbs.

Preservation of food products containing chemical food preservatives is usually based on the combined or synergistic activity of several additives, intrinsic product parameters (e.g. composition, acidity, water activity) and extrinsic factors (e.g. processing temperature, storage atmosphere and temperature).

This approach minimises undesirable changes in product properties and reduces concentration of additives and extent of processing treatments.

Lypophilic Acid Food Preservatives

Benzoic Acid

Benzoic acid in the form of its sodium salt, constitutes one of the most common chemical food preservative. Sodium benzoate is a common preservative in acid or acidified foods such as fruit juices, syrups, jams and jellies, sauerkraut, pickles, preserves, fruit cocktails, etc. Yeasts are inhibited by benzoate to a greater extent than are moulds and bacteria.

Sorbic Acid

Sorbic acid is generally considered non toxic and is metabolized; among other common food preservatives the WHO has set the highest acceptable daily intake (25 mg/kg body weight) for sorbic acid. Sorbic acid and its salts are practically tasteless and odourless in foods, when used at reasonable levels (< 0.3 %) and their antimicrobial activity is generally adequate.

Sorbates

Sorbates are used for mould and yeast inhibition in a variety of foods including fruits and vegetables, fruit juices, pickles, sauerkraut, syrups, jellies, jams, preserves, high moisture dehydrated fruits, etc.

Potassium Sorbate

Potassium sorbate, a white, fluffy powder, is very soluble in water (over 50%) and when added to acid foods it is hydrolysed to the acid form. Sodium and calcium sorbates also have preservative activities but their application is limited compared to that for the potassium salt, which is employed because of its stability, general ease of preparation and water solubility.

Gaseous Chemical Food Preservatives

Sulphur Dioxide and Sulphites

Sulphur dioxide (SO_2) has been used for many centuries as a fumigant and especially as a wine preservative. It is a colourless, suffocating, pungent-smelling, non-flammable gas and is very soluble in cold water (85 g in 100 ml at 25 °C).

The various sulphite salts contain 50-68% active sulphur dioxide. A pH dependent equilibrium is formed in water and the proportion of SO_2 ions increases with decreasing pH values. At pH values less than 4.0 the antimicrobial activity reaches its maximum. The antimicrobial action of sulphur dioxide against yeasts, molds and bacteria is selective, with some species being more resistant than others.

Sulphur dioxide and sulphites are used in the preservation of a variety of food products. In addition to wines these include dehydrated/dried fruits and vegetables, fruit juices, acid pickles, syrups, semi-processed fruit products, etc. In addition to its antimicrobial effects, sulphur dioxide is added to foods for its antioxidant and reducing properties, and to prevent enzymatic and non-enzymatic browning reactions.

Carbon Dioxide

Carbon dioxide (CO_2) is a colourless, odourless, non-combustible gas, acidic in odour and flavour. In commercial practice it is sold as a liquid under pressure (58 kg per cm^3) or solidified as dry ice.

Carbon dioxide is used as a solid (dry ice) in many countries as a means of low-temperature storage and transportation of food products. Besides keeping the temperature low, as it sublimes, the gaseous CO_2 inhibits growth of psychrotrophic micro-organisms and prevents spoilage of the food (fruits and vegetables, etc.).

Carbon dioxide is used as a direct additive in the storage of fruits and vegetables. In the controlled/ modified environment storage of fruit and vegetables, the correct combination of O_2 and CO_2 delays respiration and ripening as well as retarding mould and yeast growth. The amount of CO_2 (5-10%) is determined by factors such as nature of product, variety, climate and extent of storage.

Chlorine

The various forms of chlorine constitute the most widely used chemical sanitizer in the food industry. These chlorine forms include chlorine (Cl_2), sodium hypochlorite (NaOCl), calcium hypochlorite ($Ca(ClO)_2$) and chlorine dioxide gas (ClO_2).

These compounds are used as water adjuncts in processes such as product washing, transport, and cooling of heat-sterilised cans; in sanitising solutions for equipment surfaces, etc.

General Rules for Chemical Preservation

- Chemical food preservatives have to be used only at a dosage level which is needed for a normal preservation and not more.
- "Reconditioning" of chemical preserved food, e.g. a new addition of preservative in order to stop a microbiological deterioration already occurred is not recommended.
- The use of chemical preservatives MUST be strictly limited to those substances

which are recognized as being without harmful effects on human beings' health and are accepted by national and international standards and legislation.

Food Additive

Food additives are substances added to food to preserve flavor or improve its taste and appearance. Some additives have been used for centuries—for example, preserving food by pickling (with vinegar), or salting, as with bacon, or using sulfur dioxide, in the case of some wines. With the advent of processed foods in the second half of the twentieth century, many more additives have been introduced, of both natural and artificial origin. There is, however, an ongoing debate about the health effects of a number of these additives.

Numbering

To regulate these additives, and inform consumers, each additive is assigned a unique number. Initially, these were the "E numbers" used in Europe for all approved additives. This numbering scheme has now been adopted and extended by the Codex Alimentarius Committee to internationally identify all additives, regardless of whether they are approved for use.

E numbers are all prefixed by "E," but countries outside Europe use only the number, whether the additive is approved in Europe or not. For example, acetic acid is written as E260 on products sold in Europe, but is simply known as additive 260 in some countries. Additive 103, alkanet, is not approved for use in Europe so does not have an E number, although it is approved for use in Australia and New Zealand.

The U.S. Food and Drug Administration (FDA) listed these items as "Generally recognized as safe" or GRAS, and these are listed under both their Chemical Abstract Services number and FDA regulation listed under the U.S. Code of Federal Regulations.

Categories

Food additives can be divided into several groups, although there is some overlap between them.

Acids

Food acids are added to make flavors "sharper," and also act as preservatives and antioxidants. Common food acids include vinegar, citric acid, tartaric acid, malic acid, fumaric acid, and lactic acid.

Acidity Regulators

Acidity regulators are used to change or otherwise control the acidity and alkalinity of foods.

Anticaking Agents

Anticaking agents keep powders such as milk powder flowing freely.

Antifoaming Agents

Antifoaming agent s reduce or prevent foaming in foods.

Antioxidants

Antioxidants such as vitamin C act as preservatives by inhibiting the effects of oxygen on food, and are generally beneficial to health.

Bulking Agents

Bulking agents such as starch are additives that increase the bulk of a food without affecting its nutritional value.

Food Coloring

Colorings are added to food to replace colors lost during preparation, or to make food look more attractive.

Color Retention Agents

In contrast to colorings, color retention agents are used to preserve a food's existing color.

Emulsifiers

Emulsifiers allow water and oils to remain mixed together in an emulsion, as in mayonnaise, ice cream, and homogenized milk.

Flavors

Flavors are additives that give food a particular taste or smell, and may be derived from natural ingredients or created artificially.

Flavor Enhancers

Flavor enhancers enhance a food's existing flavors.

Flour Treatment Agents

Flour treatment agents are added to flour to improve its color or its use in baking.

Humectants

Humectants prevent foods from drying out.

Preservatives

Preservatives prevent or inhibit spoilage of food due to fungi, bacteria and other microorganisms.

Propellants

Propellants are pressurized gases used to expel food from its container.

Stabilizers

Stabilizers, thickeners and gelling agents, like agar or pectin (used in jams, for example) give foods a firmer texture. They are not true emulsifiers, but they help to stabilize emulsions.

Sweeteners

Sweeteners are added to foods for flavoring. Sweeteners other than sugar are added to keep the food energy (calories) low, or because they have beneficial effects for diabetes mellitus and tooth decay.

Thickeners

Thickeners are substances which, when added to the mixture, increase its viscosity without substantially modifying its other properties.

Food Coloring

A food coloring is any substance that is added to food or drink to change its color. It is sometimes used in cooking.

Some food colorings are extracted from natural sources, others are artificially synthesized. They are used for various purposes, such as to enhance or mask natural food colors, to provide identity to foods, and to decorate cakes and desserts. They offset the loss of natural colors when foods are exposed to light, air, temperature extremes, and moisture. Some are thought to protect flavors and vitamins present in foods from damage by light. Recent studies indicate that certain artificial coloring agents (and synthetic food preservatives) aggravate symptoms of attention-deficit hyperactivity disorder (ADHD). Several countries have therefore banned the use of some colorants.

Food coloring spreading on a thin water film.

Purpose of Food Coloring

People associate certain colors with certain flavors, and the color of food can influence the perceived flavor, in anything from candy to wine. For this reason, food manufacturers add dyes to their products. Sometimes, the aim is to simulate a color that is perceived by the consumer as natural, such as adding red coloring to glacé cherries (which would otherwise be beige). At other times, it is for effect, such as a variety of children's cereals or the green ketchup that Heinz launched in 2000.

Although most consumers are aware that foods with bright or unnatural colors likely contain food coloring, far fewer people know that seemingly "natural" foods such as oranges and salmon are sometimes also dyed to mask natural variations in color. Color variation in foods throughout the seasons and the effects of processing and storage often make color addition commercially advantageous to maintain the color expected or preferred by the consumer.

Some of the primary reasons for adding food coloring include:

- Offsetting color loss due to light, air, extremes of temperature, moisture, and storage conditions.
- Masking natural variations in color.
- Enhancing naturally occurring colors.
- Providing identity to foods.

- Protecting flavors and vitamins from damage by light.
- Decorating, such as cake icing.

Regulations

Food colorings are tested for safety by various bodies around the world and sometimes different bodies have different views on food color safety. In the United States, FD&C (Foods, Drugs and Cosmetics) numbers are given to synthetic food dyes that do not exist in nature. In the European Union, E numbers are used for all additives approved in food applications.

Most other countries have their own regulations and list of food colors, which can be used in various applications, including maximum daily intake limits.

Natural colors are not required to be tested by FDA in the United States and many other countries.

Natural Food Dyes

Several food dyes are derived from natural sources. Prominent examples are given below:

- Caramel coloring is found in cola products. It is made from caramelized sugar. It is also used in cosmetics.
- Annatto is a reddish-orange dye made from the seed of the Achiote.
- Chlorella is green and is derived from algae.
- Cochineal is a red dye derived from cochineal insects.
- Beet juice, turmeric, saffron, and paprika are also used as colorants.

Artificial Coloring in United States

Seven dyes were initially approved under the Pure Food and Drug Act of 1906, but several of those have been delisted and replacements have been found.

Current Seven

In the USA, the following seven artificial colorings are permitted in food (the most common in bold), as of 2007:

- FD&C Blue No. 1 - Brilliant Blue FCF, E133 (Blue shade).
- FD&C Blue No. 2 - Indigotine, E132 (Dark Blue shade).
- FD&C Green No. 3 - Fast Green FCF, E143 (Bluish green shade).
- FD&C Red No. 40 - Allura Red AC, E129 (Red shade).

- FD&C Red No. 3 - Erythrosine, E127 (Pink shade).
- FD&C Yellow No. 5 - Tartrazine, E102 (Yellow shade).
- FD&C Yellow No. 6 - Sunset Yellow FCF, E110 (Orange shade).

Delisted

- FD&C Red No. 2 - Amaranth (dye).
- FD&C Red No. 4.
- FD&C Red No. 32 was used to color Florida oranges.
- FD&C Orange No. 1, was one of the first water soluble dyes to be commercialized, and one of seven original food dyes allowed under the Pure Food and Drug Act of June 30, 1906.
- FD&C Orange No. 2 was used to color Florida oranges.
- FD&C Yellows No. 1, 2, 3, and 4.
- FD&C Violet No. 1.

Criticism

Although earlier research showed no correlation between ADHD and food dyes, new studies indicate that synthetic preservatives and artificial coloring agents aggravate symptoms in both those affected by this disorder and in the general population. Older studies were inconclusive quite possibly due to inadequate clinical methods of measuring offending behavior. Parental reports were more accurate indicators of the presence of additives than clinical tests. Several major studies show that academic performance increased and disciplinary problems decreased in large, non-ADHD student populations when artificial ingredients, including artificial colors, were eliminated from school food programs.

- Norway banned all products containing coal tar and coal-tar derivatives in 1978. New legislation lifted this ban in 2001 after EU regulations came into force. As such, many FD&C-approved colorings have been banned.
- Tartrazine is a coal-tar derivative, and causes hives in less than 0.01 percent of those exposed to it.
- Erythrosine is linked to thyroid tumors in rats.
- Bright food coloring ban unlikely for Australia.

Dyes and Lakes

In the United States, certifiable color additives are available for use in food as either "dyes" or "lakes".

Dyes dissolve in water, but they are not soluble in oil. They are manufactured as powders, granules, liquids, or other special purpose forms. They can be used in beverages, dry mixes, baked goods, confections, dairy products, pet foods, and a variety of other products. Dyes also have side effects which lakes do not, including the fact that large amounts of dyes ingested can color stools.

Lakes are the combination of dyes and insoluble material. Lakes tint by dispersion. They are not oilsoluble, but they are oil dispersible. Lakes are more stable than dyes and are ideal for coloring products containing fats and oils or items lacking sufficient moisture to dissolve dyes. Typical uses include coated tablets, cake and donut mixes, hard candies and chewing gums, lipsticks, soaps, shampoos, and talc.

Other Uses

Because food dyes are generally safer to use than normal artistic dyes and pigments, some artists have been using food coloring for painting pictures, especially in forms such as bodypainting. Also, food coloring can serve as a means of dyeing fabric. It can be fixed on nylon and animal fibers, but it is not wash-fast when used on cotton, hemp, and other plant fibers.

References

- What-is-the-importance-of-food-preservation-5187: preservearticles.com, Retrieved 20 April 2018
- Food-microbiology, low-temperature-storage-chiling-and-freezing-59275: biologydiscussion.com, Retrieved 14 May 2018
- Non-thermal-plasma-as-a-new-food-preservation-method-Its-present-and-future-prospect-259622176: researchgate.net, Retrieved 11 June 2018
- Water-Activity-Concept-and-its-Role-in-Food-Preservation-310605656: researchgate.net, Retrieved 30 July 2018
- Food-preservation, Dehydration-50570: britannica.com, Retrieved 15 may 2018
- Food-coloring: newworldencyclopedia.org, Retrieved 10 June 2018

4

Food Packaging

Food packaging is the process of enclosing food to prevent it from damage, contamination, spillage and spoilage. This chapter delves into various aspects of food packaging such as active packaging, intelligent packaging, packaging machines, symbols of packages, materials used in packaging, etc. to provide an in-depth understanding of the subject.

Food packaging is the enclosing of food to protect it from damage, contamination, spoilage, pest attacks, and tampering, during transport, storage, and retail sale. The package is often labeled with information such as amount of the contents, ingredients, nutritional content, cooking instructions (if relevant), and shelf life. The package needs to be designed and selected in such a manner that there are no adverse interactions between it and the food. Packaging types include bags, bottles, cans, cartons, and trays.

Automated palletizer of bread with industrial robots at a bakery.

Functions of Food Packaging

Food packaging serves many important functions. They may be broken down as follows.

- Containment: For items that are granulated, paper-based packages are the best, with a sealing system to prevent infiltration of moisture into the product. Other products are packaged using metal cans, plastic bags and bottles, and glass containers. Another factor in containment is packaging durability—in other words, the packaged food has to survive transport from the food processing facility to the supermarket to the home for the consumer.

- Protection: The packaging must protect the food from (a) biological agents such as rats, insects, and microbes; (b) mechanical damage such as product abrasion, compressive forces, and vibration; and (c) from chemical degradation such as oxidation, moisture transfer, and ultravioletlight.
- Communication: Packaged food must be identified for consumer use, mainly with label text and graphics. It can also be done by using special shapes for the food package, such as the Coca-Cola bottle or the can of Spam. Other well-known food package shapes include potato chip bags and milk bottles. These packages also detail nutritional information, and whether they are packaged according to kosher or halal specifications. The label may also indicate whether it is safe to put the packaged food (such as a TV dinner) through a microwave process.
- Environmental issues: To protect the environment, we must be willing to reuse or recycle the packaging or reduce the size of the packaging.
- Package safety: Before using a particular type of package for food, researchers must ensure that it is safe to use that packaging for the food being considered, and that there are no adverse interactions between the package and the food. This includes any metal contamination issues from a can to the food product or any plastic contamination from a bottle to the food product.
- Product access: The packaging must be such that the product is readily accessible when the consumer is ready to use it. For example, pour spouts on milk cartons can make it easy to dispense the milk.

Food Packaging Types

The materials mentioned above can be fashioned into different types of food packages and containers. Examples are given below.

Packaging type	Type of container	Examples of foods packaged
Aseptic packages	Primary	Liquid whole eggs
Plastic trays	Primary	Portion of fish
Bags	Primary	Potato chips
Bottles	Primary	Bottle of a soft drink
Boxes	Secondary	Box of soft drink bottles
Cans	Primary	Can of tomato soup
Cartons	Primary	Carton of eggs
Flexible packaging	Primary	Bagged salad
Pallets	Tertiary	A series of boxes on a single pallet, to transport packaged food from the manufacturing plant to a distribution center.
Wrappers	Tertiary	Used to wrap the boxes on the pallet for transport.

Primary packaging is the main packaging that holds the food that is being processed. Secondary packaging combines the primary packages into a single box. Tertiary packaging combines all of the secondary packages into one pallet.

Special Techniques

- Vacuum packaging or inert atmosphere packaging: Oxygen in the air tends to reduce the shelf life of food by the process known as oxidation. To prevent this process, some foods are packaged at reduced pressure (partial vacuum) or using an inert gas (such as nitrogen) to replace oxygen.
- Bags-in-boxes: These are used for soft drink syrups, other liquid products, and meat products.
- Wine box: This is a type of box designed for storage of wine.

Packaging Machines

The design and use of packaging machinery needs to take into account the following factors: technical capabilities, labor requirements, worker safety, maintainability, serviceability, reliability, ability to integrate into the packaging line, capital cost, floorspace, flexibility of use, energy usage, quality of outgoing packages, qualifications (for food, phamaceuticals, and so forth), throughput, efficiency, productivity, and ergonomics.

Packaging machines may be of the following general types:

- Blister, Skin and Vacuum Packaging Machines.
- Capping, Over-Capping, Lidding, Closing, Seaming and Sealing Machines.
- Cartoning Machines.
- Case and Tray Forming, Packing, Unpacking, Closing and Sealing Machines.
- Check weighing machines.
- Cleaning, Sterilizing, Cooling and Drying Machines.
- Conveying, Accumulating, and Related Machines.
- Feeding, Orienting, Placing, and Related Machines.
- Filling Machines: handling liquid and powdered products.
- Package Filling and Closing Machines.
- Form, Fill and Seal Machines.
- Inspecting, Detecting and Checkweighing Machines.

- Palletizing, Depalletizing, Pallet Unitizing and Related Machines.
- Product Identification: for labeling, marking, and so forth.
- Wrapping Machines.
- Converting Machines.
- Other specialty machinery.

Active Packaging

Active packaging is an innovative approach to maintain or prolong the shelf-life of food products while ensuring their quality, safety, and integrity.

Methods of Active Packaging

Oxygen Scavengers

Packaged foods include a certain amount of headspace gases and entrained oxygen. Permeation of oxygen into plastic containers is also of concern. Molecular oxygen (O_2) can be reduced to a variety of intermediate species by the addition of between one and four electrons thereby forming superoxide, hydroxy radical, hydrogen peroxide, and water. Except water, the remaining three intermediate species are very reactive. These reactive oxygen species are free radical in nature so the oxidative reactions in in which they participate are therefore autocatalytic. The presence of oxygen, if not desired, may result from inadequate or insufficient evacuation during the packaging process, presence in the food itself or the packaging material and release into the headspace, permeation through the package, introduction of air due to a poor sealing, or micro perforations in the packaging material. A high level of oxygen reduces the nutritional value of food and reduces its shelf life. The oxygen in headspace gases react with sensitive foods in the package and accelerates the deterioration of many food products (meats, sausages, milk powder, or spices), degradation of vitamins, and rancidity of oils, nuts, and fatty foods, and also encourages microbial growth. Oxygen in the headspace of food packaging can be removed by vacuum sealing or by inert gas atmosphere in the packaging (N_2, CO_2), or both. These technologies can remove about 90–95% of the oxygen present in air from the packed food prior to or during packaging. Such systems are used in packaging orange juice and in the brewing industries, and in modified-atmosphere packaging of food products. It is imperative to remove oxygen during packaging, control of the residual oxygen level in the package by use of oxygen absorbent materials limits the rate of deterioration and food spoilage. In fresh meat, the presence of oxygen allows oxygenation of myoglobin, which imparts the characteristic red color; since consumers judge meat by its appearance, texture, and flavor, this characteristic of meat is important because. However, high levels of oxygen promote the oxidation of muscle lipids, which eventually has detrimental effects on the color of fresh meat The shelf-life of meat is increased by decreasing oxygen levels, as this prevents growth of fungi and aerobic bacteria Oxygen scavengers reduce and actively control the residual levels of oxygen inside the package, in some cases to <0.01% oxygen, which is impossible with other packaging systems.

- They prevent oxidation phenomena: rancidification of fats and oils and consequent emergence of off-odors and off-flavors, loss or change of colors characteristic of food, loss of oxygen-sensitive nutrients (vitamins A, C, E, unsaturated fatty acids, etc.).
- They prevent the growth of aerobic microorganisms.
- They reduce or eliminate the need for preservatives and antioxidants in food by incorporating the added value of "fresh" or "natural".
- They are an economical and efficient alternative to the use of modified atmosphere and vacuum packaging.
- They slow down metabolism of food.

The use of these systems, either alone or in combination with other traditional packaging systems, and the use of modified atmospheres can therefore extend the commercial life of a food product.

Different mechanisms of action of oxygen scavengers are:

Oxidation of iron and iron salts: This is the most widely used most effective. Oxygen scavenger systems that are based on iron oxidation reactions are explained by the following equation:

$$4\,Fe(OH)_2 + O_2 + 2\,H_2O \rightarrow 4\,Fe(OH)_3$$

This system is based on the oxidation of iron and ferrous salts (provided in the packet) that react with water provided by food to produce a reaction that moisturizes the iron metal in the product packaging and irreversibly converts it to a stable oxide. The iron powder is contained within small oxygen permeable bags that prevent contact with food.

- Oxidation of ascorbic acid and unsaturated fatty acids (oleic, linoleic).
- Oxidation of photosensitive coloring matter.
- Enzymatic oxidation by Glucose oxidase.

These methods can reduce oxygen levels to < 0.01%, which is much lower than the typical level of 0.3–3.0% obtained with residual oxygen-modified atmosphere packaging (Sen et al 2012). This lower level of oxygen can be maintained for long periods depending upon the oxygen permeability of the packaging material.

The disadvantages of sachets is the need for additional packaging operations to add the sachet to each package, also they cannot be used in beverages or foods containing high levels of water because they become inactive when wet. Also in high moisture foods aqueous slurry of oxygen absorbent is formed when moisture enters in the adsorbent packet. The aqueous slurry oozes out of the package on to the foodstuff, spoiling its appearance. In contrast, ascorbic acid can be used in liquid food or beverage as an efficient oxygen scavenger.

Commercial presentations of oxygen scavengers are:

- Independent systems such as bags, strips, or labels, which are incorporated into or attached to the inside of the package, but are separate elements. They are the most widely used systems.
- Systems integrated into the packaging material itself, not visually perceptible as distinct elements. Iron, ascorbic acid, and low molecular weight ingredients are used in this way.
- Selection of oxygen scavenger (shape, size, ability to absorb oxygen, time to reach equilibrium) must be very strict and tailored to the needs and characteristics of each food (liquid, solid, dry, fat, water content, activity water, etc.) and storage temperature. Fruits and vegetables, seafood, cheeses, baked goods, pastries, cookies, pizza crusts, and pasta can benefit from packaging systems with both oxygen absorption and antimicrobial release systems.

Carbon Dioxide Generating System

Carbon dioxide suppresses microbial activity. Relatively high CO_2 levels (60 to 80%) inhibit microbial growth on surfaces and, in turn, prolong shelf life of packed food. Therefore, a complementary approach to O_2 scavenging is the impregnation of a packaging structure with a CO_2 generating system or the addition of the latter in the form of a sachet. Since the permeability of CO_2 is 3 to 5 times higher than that of O_2 in most plastic films, it must be continuously produced to maintain the desired concentration within the package. High CO_2 levels cause changes in taste of products so a CO_2 generator is only useful in certain applications such as fresh meat, poultry, fish and cheese packaging. oxygen-free environment alone is insufficient to retard the growth of Staphylococcus aureus, Vibrio species, Escherichia coli, Bacillus cereus and Enterococcus faecalis at ambient temperatures. O_2 and CO_2 absorber inhibited the growth of Clostridium sporogenes.

Ethylene Scavengers

The control of ethylene in stored conditions plays a key role in prolonging the postharvest life of many types of fresh produce. Most fruits and vegetables release ethylene after they are harvested. Ethylene is a phytohormone that initiates and accelerates ripening, produces softening and degradation of chlorophylls, and inevitably leads to deterioration of fresh or minimally processed fruits and vegetables. Ethylene scavengers are useful for preserving ethylene-sensitive fruits and vegetables such as apples, bananas, mangos, tomatoes, onions, carrots.

Mechanism of Action

- One of the main mechanisms of action of ethylene scavengers is based on the use of potassium permanganate, which oxidizes ethylene to carbon dioxide and water. The typical permanganate content is between 4%and 6%. Potassium permanganate oxidizes ethylene and changes color from purple to brown, and thus, a color change indicates its residual ethylene absorbing capacity, but because of its toxicity potassium permanganate cannot be used in direct contact with food.
- Other systems are based on the ability of certain materials to absorb ethylene, alone or with any oxidizing agent. For example, palladium has been shown to have a higher ethylene adsorption capacity than permanganate-based scavengers in situations of high relative humidity.

LDPE and HDPE polyethylene films as packaging material are able to absorb ethylene; ethanol, ethyl acetate, ammonia, and hydrogen sulfide are used in food industry. These films keep food fresh for longer and eliminate odors.

Flavor and Odor Absorber/Releaser

Addition of essences and odors can increase the desirability of the food to the consumer, to improve the aroma of fresh product itself, or to enhance the flavor of food when the package is opened. These flavors and aromas are released slowly and evenly in the packaged product during its shelf life or release can be controlled to occur during opening the package or food preparation. Gradual release of odors can offset the natural loss of taste or smell of products with long shelf lives.

Antimicrobials

Microbiological contamination due to pathogenic or spoilage bacteria may occur during inadequate processing, or when package integrity is compromised due to a ruptured seal, puncture, dents, or incomplete glass finishes. Traditional methods of preserving food from the damaging effects of microbial growth include heat treatment, drying, freezing, refrigeration, irradiation, modified-atmosphere packaging, and addition of salts or antimicrobial agents. Antimicrobial packaging include systems such as adding a sachet into the package, dispersing bioactive agents in the packaging, coating bioactive agents on the surface of the packaging material, or utilizing antimicrobial macromolecules with film-forming properties or edible matrices. A large number of agents with antimicrobial properties (ethanol, carbon dioxide, silver ions, chlorine dioxide, antibiotics, organic acids, essential oils and spices, etc.) are being used for the purpose of inhibiting the growth of microorganisms that can lead to deterioration of foodstuffs (bacteria can also attack the packages affecting their functions and properties).

Packaging systems that release volatile antimicrobials also include chlorine dioxide, plant extracts, sulfur dioxide, essential oils, carbon dioxide, and allylisothiocyanate release systems. The theoretical advantage of volatile antimicrobials is that they can penetrate most of the food matrix and the polymer not necessarily in direct contact with food. This type of active packaging is suitable for applications where contact between the portions of food and packaging does not occur, as in the case of ground beef.

Chlorine dioxide can exist in gaseous, liquid, or solid form. It has proven effective not only against bacteria and fungi but also against viruses. Potential applications of chlorine dioxide include meat, poultry, fish, dairy products, and confectionery and baked goods.

Sulfur dioxide is the most effective material for controlling decomposition of grapes and is much more effective than the combination of γ-radiation and heat. However, it has drawbacks that include bleaching of grape skin and the fact that some sulfur dioxide may remain on grapes.

The role of carbon dioxide in the packaging atmosphere is to suppress microbial growth and to slow down the rate of respiration of fruits and vegetables. Because the permeability of carbon dioxide is between 3 and 5 times that of oxygen in most packaging films, carbon dioxide must be continuously released to maintain the desired concentration in the package (Ozdemir and Floros 2004) High levels of carbon dioxide (10–80%) are suitable for foods such as meat and poultry in order to inhibit microbial growth on the surface of the products and to extend the shelf life.

Another way of tackling the problem of microbial growth is by the use of nonvolatile antimicrobial additives. Many preservatives (sorbic acid, benzoic acid, propionic acid and its salts, or bacteriocins such as nisin, natural spices, silver ions, chelators, etc.) are added to plastic films and materials used as antimicrobials. But these nonvolatile antibacterial require direct contact with the food to be active.

Antioxidants

Oxidation of fats is one of the most important mechanisms leading to food spoilage, second only to growth of microorganisms. The oxidation of lipids in food leads to a reduction in shelf-life due to changes in taste and odor, deterioration of the texture and functionality of muscle foods, and a reduction in nutritional quality. Oxidation of food can be avoided by use of oxygen scavengers and antioxidant agents in the packaging. Such packaging is intended to prevent or slow down the oxidation reactions that affect the quality of food. However, radicals, mainly oxo, hydroxyl, and superoxide, are originated from oxygen and they are the main initiators of oxidation. Thus, oxidation can be avoided by eliminating radicals as soon as they are formed.

A varnish with natural antioxidant of rosemary extract, which acts as a radical scavenger either in the vapor phase or by direct contact, can eliminate or delay the oxidation of foods inside the food package. This eliminates the need to add antioxidants to the package or the food The use of antioxidant active film in the conservation of fresh meat can enhance the stability of myoglobin and fresh meat against oxidation processes. Migration of α tocopherol from a multilayer active packaging (made of high-density polyethylene, ethylene vinyl alcohol, and a layer of low-density polyethylene containing the antioxidant α-tocopherol) shows a delay in lipid oxidation in whole milk powder.

The antioxidant content decreases during storage due to diffusion of the antioxidant through the film and its subsequent evaporation at the surface. This decrease in the concentration of the antioxidant can be prevented by adding an extra layer of film. Antioxidants can be used for oil, nuts, butter, fresh meat, meat derivatives, bakery products, fruits and vegetables.

Intelligent Packaging

The headspace of food packages undergoes changes in their composition over time. Devices capable for identifying, quantifying, and reporting changes in the atmosphere within the package, the temperatures during transfer and storage and the microbiological quality of food are the basis of intelligent packaging. The indicators should be easily activated and exhibit a change (or show an indication) that is easily measurable and irreversible, time- and temperature-dependent changes must be reproducible and ideally matched or readily correlated with the food quality, and also provide information regarding the status of the package.

Time-temperature Indicators

They are of two types: Visual indicators and radio frequency identification (RFID) tag. The visual indicators change color in response to cumulative exposure to temperature. The main mechanisms of action include enzymatic reactions, polymerization, or chemical diffusion. These products are used to monitor exposure to unsuitable temperatures during transport and storage and are an indication of quality for the producer because they ensure that the product reaches the consumer in optimal conditions.

The RFID tag is an advanced form of data carrier for automatic product identification and traceability. In an RFID system, a reader emits radio waves to capture data from an RFID tag, and the data are then passed to a host computer for analysis and decision making. The RFID tag contains a minuscule microchip connected to a tiny antenna.

Seal and Leak Indicators

The gas composition in the package headspace often changes as a result of the activity of the food product, leaks, nature of the package, or environmental conditions. O_2 and CO_2 can be used to monitor food quality, as seal indicators (leaks), or to verify the effectiveness of an oxygen absorber. Most O_2 or CO_2 indicators change color as a result of chemical or enzymatic reactions. A color change indicates when the oxygen concentration exceeds the limit established in a sealed food package.

A major problem with such indicators is that they require storage under anaerobic conditions, since they quickly deteriorate in air.

Freshness and Ripening Indicators

Freshness and ripening indicators provide an indication of the deterioration or loss of freshness of packaged goods. They are described as indicating different mechanisms of volatile metabolites, such as diacetyl, amines, carbon dioxide, ammonia and hydrogen sulfide, produced during the aging of foods.

Changes in the concentration of hydrogen sulfide or organic acids such as n-butyrate, L-lactic, D-lactate, and acetic acid during storage are offered as viable indicators of the formation of metabolites in meat products, fruits, and vegetables. Indicators based on color changes due to changes in pH are of great potential use as indicators of microbial metabolites and ripeness.

Products formed during microbial growth (carbon dioxide and hydrogen sulfide) and biogenic amines are of great potential use in indicating the freshness of meat and fish. Biogenic amines (putrescine, cadaverine, histamine, and others) are formed by degradation of protein-containing food to amino. Thus, biogenic amines are an indicator of food deterioration and only an indirect indicator of food freshness in meat and fish.

Packaging separates the products from the external environment and has in general four basic functions: protection, communication, convenience, and containment. It communicates with the consumer by written texts or graphics and simplifies the handling of the contained products with practical features such as reclose- or microwave-ability. Furthermore, it offers differently shaped and sized containers and adapts to the customer's lifestyle. In addition to improved marketing and distribution, packaging also slows quality decay. That is why they contribute significantly to safe delivery and preservation of packaged food. However, a complete elimination of the quality loss is not possible. Intrinsic properties of highly perishable foods change after processing. This can lead to an increase in quality (e.g., the ripening of fruits to a certain level) or to a quality loss: Depending on the contents of the package, biological, chemical, or physical processes occur, which ultimately lead to spoilage of the product. These changes are in most cases difficult to assess by the consumers. In fear groceries could be spoiled, many consumers throw products away which would actually have still been suitable for consumption. Often a small deviation from the norm, either the color, the consistency, or even the passing of the best before date leads to products ending up in the bin.

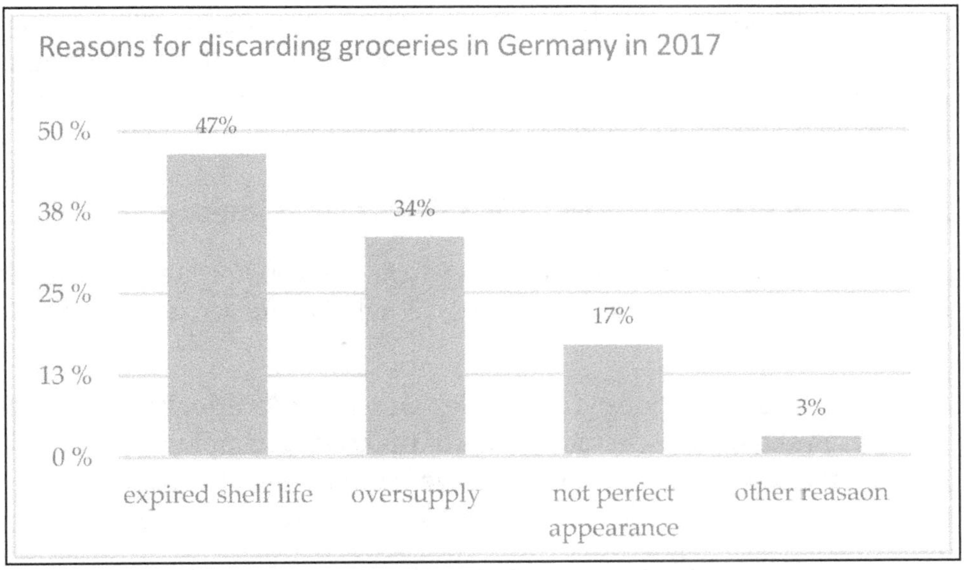

Reasons for discarding groceries in Germany in 2017.

If we compare the loss of food within the value chain, it is the end consumer who contributes most. What is even worse is that much of the food loss could be avoided.

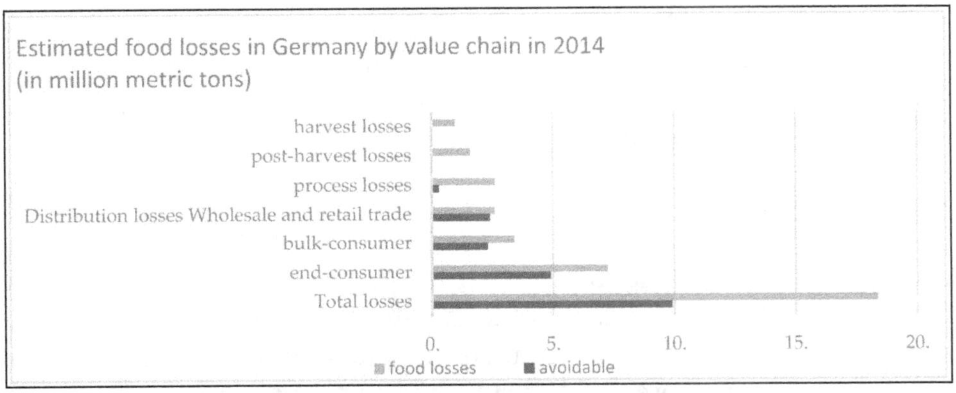

Estimated food losses in Germany by value chain in 2014.

To reduce this unintentional product waste, so-called intelligent packaging concepts could be utilized. But this is not the only advantage these technologies provoke.

Microbiological and chemical tests of the products are regularly performed at the company level during production and before delivery. But in most cases there is no such control after delivery to the supermarket. Intelligent packaging will close this gap as they are able to monitor and display the quality status from the point of manufacture up to the customer. This permanent monitoring not only minimizes unnecessary food waste but also protects consumers against potential food poisoning, maximizes the efficiency of the food industries, and improves traceability.

Preserving food quality is also an important area of research, as it is directly related to the global aim of improving the quality of our lives. Furthermore, there is a growing demand by consumers due quality and safety properties. These issues are highly dependent on the applied packaging materials. Intelligent Packaging has potential to improve product safety, reduce environmental impact and increase the attractiveness of the packaged product and also of the food companies.

Different Types and Concepts of Intelligent Packaging

In general there are three main technologies which are used for intelligent packaging systems: data carriers, indicators, and sensors. A subdivision according to the following types is also possible:

- Environmental conditions: This species monitors conditions which can lead to changes in the quality characteristics of the food. Examples of these types are time temperature indicators, gas leakage indicators, and relative humidity sensors. Depending on the monitoring factor, these systems can be placed outside or inside the packaging.
- Quality characteristics or quality indicator compounds: This type is used for the direct monitoring of the quality attributes of the food itself. Examples are bio sensors and freshness sensor/indicators. These devices are usually located inside the packaging.
- Data carriers: These systems are only used to store and transfer data, while indicators and sensors are used to monitor the external environment and display the information afterwards.

Data Carriers

Data carriers help to make the information flow within the supply chain more efficient. The function of data carriers is not to monitor the quality of the products, but to guarantee traceability, automatization, theft protection, or counterfeit protection. To ensure this, data carriers store and transmit information about storage, distribution, and other parameters. Therefore, they are often placed on tertiary packaging. The most frequently used data carriers are barcode labels and RFID (Radio Frequency Identification) tags.

Barcodes

Barcodes are cheap, easy to use, and are widely used to facilitate inventory control, stock recording, and checkout. In general, barcodes can be divided into one-dimensional and two-dimensional ones. Depending on the type, they have different storage capacities.

A one-dimensional barcode is a pattern of parallel spaces and bars. The different arrangement of the bars and gaps results in the coding of data. A barcode scanner and an associated system can translate the coded information.

Barcode.

Two-dimensional barcodes offer more memory capacity (e.g., for packaging date, batch number, packaging weight, nutritional information, or preparation instructions) because of the combination of dots and spaces arranged in an array or a matrix. This provides great convenience for retailers and consumers. An example of 2D barcodes are QR (quick response) Codes.

QR-Code.

Radio Frequency Identification (RFID) Technology

RFID tags are advanced data carrier with a data storage up to 1 MB, as well as non-contact and non-line-of-sight ability to collect real time data. These collect, store and transmit real-time information to a user's information system. In comparison to barcodes, RFID tags are more expensive and need a more powerful electronic information network. On the other side, the information can be loaded electronically on these tags and can be changed again. Furthermore, RFID offers further advantages for the entire food supply chain. These include traceability, inventory management and promotion of quality and safety.

A RFID system consists of three compounds: A tag formed by a microchip connected to a tiny antenna, a reader that emits radio signals and receives answers from the tag in return and a middleware that bridges the RFID hardware and enterprise applications.

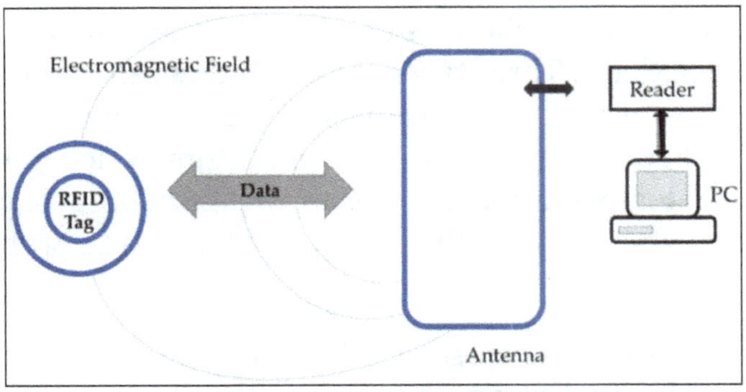

The working principle of radio frequency identification (RFID) tag.

Time Temperature Indicator (TTI) Integrated Barcodes and RFID Tags

Barcodes and QR Codes are the first so-called intelligent packaging technologies. Meanwhile they have been further developed and integrated into TTIs. The principle is based on the fact that a label is scanned and information about the product as well as the temperature progression is given. Compared to traditional data carriers, these systems can not only be used to track the distribution chain, but can also help reduce food waste.

For example, Bioett has a TTI-Barcode system on the market where the data is captured with a portable scanner, displayed on a computer monitor and downloaded to a database for analysis. Infratab has developed a battery powered TTI-RFID tag that uses a microchip to capture the temperature progression to determine the shelf life of a product. A biosensor-barcode, called Food Sentinel System, was developed by SIRA Technologies. A specific pathogenic antibody is bonded to a membrane-forming part of the barcode. If bacteria are present, a dark bar is formed that makes the barcode unreadable when scanning.

Indicators

Indicators determine the presence or absence of a substance, the extent of a reaction between different substances or the concentration of a particular substance. This information is visualized by direct changes, e.g., different color intensities. Depending on the indicator they are placed inside or outside of the package.

Time Temperature Indicators (TTIs)

Temperature is an important factor in determining the shelf life of a food product. Deviations in the temperature profile can result in growth or survival of microorganisms, which ultimately causes in spoilage of the product. Furthermore, incorrect freezing can denature the proteins of meat or other products. Whether the cold chain or a required temperature is properly maintained during the food supply chain, time temperature indicators can be used.

In general, time temperature indicators or integrators are simple, inexpensive gadgets attached to the package. Three types can be distinguished: Critical temperature indicators, which show if products have been heated above or cooled below a permitted temperature. Secondly, partial history indicators, which indicate if a product has been subjected to temperature, that cause a change in product quality. Thirdly, a full history indicator which records the complete temperature profile along the food supply chain.

The functional principle of TTIs is based on the detection of time and temperature dependent mechanical, chemical, electrochemical, enzymatic, or microbiological changes of a food product. For example, chemical or physical responses are based on acid–base reactions or polymerization towards time and temperature. In contrast, biological responses are based on biological changes such as microorganisms, spores or enzymes in relation to time and temperature. The measured values are usually expressed as a visible response, like color changes or mechanical deformations.

Because of this simple functionality TTIs are recognized as user-friendly and readily usable devices. An example of a TTI indicator is the Fresh-Check from Lifeline technologies. Its function is based on a polymerization reaction resulting in a color change in the indication range. A clear center indicates a fresh TTI. If the color of the active center matches the outer ring, the product should be consumed soon. TTIs of not fresh products have a dark center.

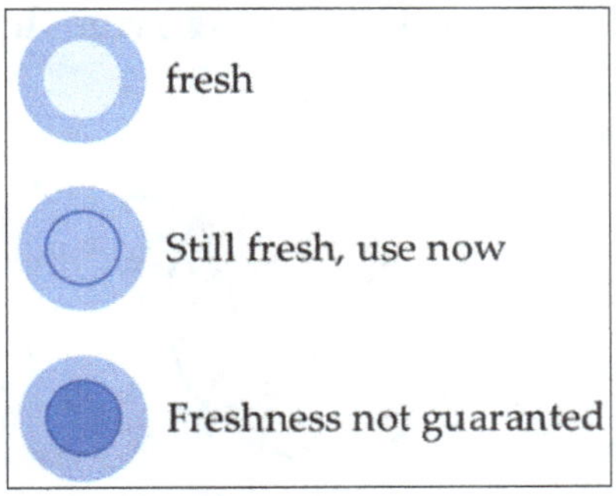

Principle of Lifeline's Fresh-check Indicator time temperature indicator (TTI).

Freshness Indicators

Freshness indicators supervise the quality of food products during storage and transportation. Reasons for the loss of freshness can be disadvantageous conditions or exceeded durability. Therefore, they submit information about microbiological growth, presence of microbiological metabolites or chemical changes of the products. Quality indicating metabolites are for instance glucose, organic acids, ethanol, volatile nitrogen compounds, biogenic amines, carbon dioxide, ATP degradation products, and sulphuric compounds. To be able to be in contact with the compounds the freshness indicators must be placed inside the packaging. Depending on the indicator, this information can be detected by different methods.

Table: Principles of indicators and sensors based on metabolites (based on).

Metabolites	Food Products	Indicators	Sensor
Glucose/lactic acid	Fermented food, meat.	Colorimeter based on pH.	Electrochemical sensor by redox reaction.
Carbon dioxide	Fermented food, meat, seafood.	Colorimeter based on pH.	Electrochemical sensor by silicon-based polymers.

Oxygen	Meat, vegetable, fruits.	Optical sensor by fluorescence, colorimeter based on pH.	Electrochemical sensor, laser.
Biogenic amines	Fish, meat.	Color-changing pH-sensitive dyes.	Electrochemical sensor by enzyme redox reaction.

An example for a freshness indictor is a sensor label from FQSI (Food Quality Sensor International Inc., Lexington, MA, USA), which is able to detect biogenic amines. The SensorQ™ sticker ticket is applied on the inside of the packaging and indicates that a critical level of bacterial growth has been reached by a change of color (orange to brown). The operating principle of biogenic amine sensors is based on amine oxidases or transglutamase. Lactic acid sensors operate on the basis of lactate oxidase and peroxidase activities. Glucose sensors use glucose oxidases that are immobilized on the surface of the electrodes. Glucose oxidase is an enzyme that catalyzes the oxidation of glucose.

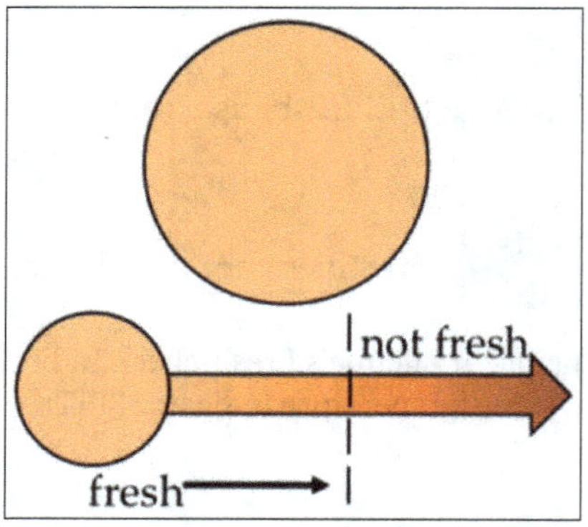

Principle of SensorQ, Smart Sensor Label by FQSI.

Gas Indicators

Gas indicators indicate the quality condition of the food depending on the indoor atmosphere. A sensor detects and reacts to changes in the atmosphere inside the packaging, while the actual indicator displays the quality status. The modifications of the atmosphere are based on the one hand of the food activity, such as enzymatic or chemical reactions, and on the other hand of the package nature and environmental conditions, such as gas generation by microorganisms metabolism or gas transmission through the packaging. Most of them monitor the oxygen and carbon dioxide concentrations. But also water vapor, ethanol, hydrogen sulfide, and other gases are checked. The concentrations of these gases often correlate closely with the advance of spoilage. The functionality of most devices is based on redox dyes, a reducing compound and an alkaline component. In order to be able to monitor the gases, indicators must be placed inside the packaging. But many of these indicators have a loss of color due to the moisture in the packaging. However, companies are already researching UV-activated colorimetric indicators that show less dye leaching because of encapsulation or coating technologies.

Sensors

A sensor is defined as a device used to detect, locate or quantify energy or matter giving a signal for the detection or measurement of a physical or chemical property to which the device responds.

Most sensors consist of two components: They have a sensor part which is also known as the receptor. This can detect the presence, activity, composition or concentration of certain chemical or physical analytes. The physical or chemical information is also converted by the receptor into a form of energy that can be measured by the second component, the transducer. Furthermore, the transducer is used to convert the measured signal into a useful analytic signal. This can be an electrical, chemical, optical or thermal signal.

There are different types of sensors that investigate different parameters, for instance the gas sensors. The progress of spoilage can be determined by the concentration of certain gases, like CO_2 or H_2S. The gas sensors make use of these properties by monitoring them. Those respond quantitatively and reversibly to the presence of a gas by changing the physical parameters of the sensor. CO_2 sensors are mostly non-dispersive infrared (NDIR) sensors or chemical sensors. NDIR sensors are spectroscopic sensors that measure the CO_2 content by gas absorption at a certain wavelength. Chemical CO_2 sensors work with polymer or solid electrolytes. Infrared sensors as well as electrochemical, ultrasonic and laser technologies are used for the detection of O_2.

Another type of sensors are the biosensors. Compared to chemical sensors, they have a receptor made of biological materials such as enzymes, antigens, hormones, or nucleic acids. Depending on the measuring parameters, the transducer can be electrochemical, optical, acoustic, etc. For example, there is a biosensor (Toxin Guard by Toxin Alert) whose functional system is based on antibodies, which are integrated in plastic packaging and thus make it possible to detect pathogens such as *Salmonella*, *E. coli*, *Listeria*, and *Campylobacter*. A positive result is indicated by a visual signal. Another biosensor is able to detect xanthine, which is an adenine nucleotide degradation product in animal tissue. To do this, xanthine oxide is immobilized on platinum, silver or pencil graphite electrodes.

Advantages and Disadvantages of Intelligent Packaging Concepts

In general, intelligent packaging is easy to use and provide a number of advantages for consumers, food manufacturers, and the whole food industry. Depending on the system they offer different features.

The current quality status of a product can be determined by the use of indicators and sensors. This results in a general increase of product safety and in a reduction of unnecessary food waste. In addition, this consistent quality monitoring also reduces time and material costs in the analysis methods of packaged food. Further cost advantages also arise along the supply chain when intelligent packaging minimizes food waste. These aspects could be even more important in other life sciences like the pharmaceutical industry.

Data Carriers enable better traceability of the supply chain. Because of their low price, ease of use, and the benefit they provide, barcodes and QR Codes are nowadays widely spread. In contrast, indicators and sensors can be barely found on the market. One reason for this is the price as the development and production costs are still very high. The packaging costs can amount to 50–100% of the total costs of the final product. Actually a limit for packaging costs of 10% of the products value to be packaged is provided. Furthermore, the use of indicators and sensors could lead to a negative change in consumer buying behavior: Customers would most likely put products with a discolored freshness indicator back on the shelf and choose a product with an uncolored freshness indicator. If the customer often sees labels of a brand product with a divergent color, he could even loose his confidence in that brand. At the same time, this behavior could also lead to an increase in unsold foodstuff. On the other side, intelligent packaging can optimize the classic "first in—first out" principle. As the real current quality status of the food is known, the retailer can sell the products with the shorter shelf life first and so the wastage of food could be reduced.

It must be ensured that the systems are compatible with the food to be monitored. Not every intelligent packaging can be used for each type of food. Therefore, it must be clarified which indicator or sensor is appropriate for the product. The intelligent packaging can only be advantageous if it matches with the food. For instance, an oxygen sensor would be useful for MAP (Modified Atmosphere Packaging) packaged foods, while for chilled and frozen products a TTI should be apply.

Another aspect that still needs to be clarified is the recycling of the packaging. The additional waste generated by the installation and production of intelligent packaging is actually contradictory to the goal of reducing the amount of food wastage.

It should also be noted that it is not possible to rely a 100% on intelligent packaging for optimum product quality as misuse or failure of the systems cannot be ruled out. Several factors are often responsible for the loss of quality. Monitoring just one parameter cannot provide a complete statement about the quality status of a product. Furthermore, external environmental influences such as light, temperature or mechanical stress can sometimes have an adverse effect on the technologies. On the one hand this can lead to a situation, where products are classified as no longer fit for consumption, even though they still are. On the other hand, this can result in a situation in which the spoilage of a product is not indicated. In the worst case, the consumer's health may be adversely affected if the products are consumed. To sum up it can be said that the robustness of the systems must be improved and the individual packaging technologies should be combined in order to exploit as many advantages as possible.

Types of Intelligent Packaging

Intelligent Packaging: Time Temperature Indicators (TTIs)

The intelligent packaging design that is leading the way in packaging technology is the time temperature indicator (TTI). The TTI is useful because it can tell the consumer when foods have been temperature abused. If a food is exposed to a higher temperature recommended, the quality of the food can deteriorate much quicker. A TTI can be placed on shipping containers or individual packages as a small selfadhesive label, and an irreversible change, like a color change, will result when the TTI experiences abusive conditions. TTIs are particularly useful with chilled or frozen foods, where the cold storage during transportation and distribution are important for food quality and safety. TTIs are also used as freshness indicators for estimating the shelf life of perishable products. A TTI technology known as Timestrip is currently being employed by Nestle in their food service products in the UK. The Timestrip uses a steady diffusion of liquid through a membrane to measure the time that has elapsed at a particular temperature. This action can provide information about how long a product has been opened or in use. The Timestrip is very useful for products like sauces that have to be refrigerated and used within a specific time period.

Intelligent Packaging: Gas Indicators

Food is a complicated material to package because it is capable of respiration and therefore may change its own atmosphere when inside a package. The gas composition within a package can easily change due to the interaction of food with its environment. Gas indicators are a helpful means of monitoring the composition of gases inside a package by producing a change in the color of the indicator though a chemical or enzymatic reaction. The indicators must be in direct contact with the gaseous environment directly surrounding the food in a package. Indicators are capable of signaling whether there is a gas leakage in the package, or they may be used to verify the efficiency of an oxygen scavenger. Gas indicators typically signal the presence or absence of oxygen and carbon dioxide. Oxygen in the air can cause oxidative rancidity, unwanted color changes in foods, and allow aerobic microbes to grow on foods. Oxygen indicators typically result in a color change when oxygen is present, and the presence of oxygen can indicate that the package has a leak or has been tampered with. Oxygen indicators can also indicate improper sealing of a package. Gas indicators are also being developed to detect water vapor, ethanol, and hydrogen sulfide.

Intelligent Packaging: Thermochromic Inks

Inks are available that are temperature sensitive and can change colors based on temperature. These inks can be printed onto packages or labels such that a message can be conveyed to the consumer based on the color of the ink they are seeing. Thermochromic inks can let a consumer know whether a package is too hot to touch, or cold enough drink. Thermochromic inks are becoming a popular technology for beverages. The inks used can be adversely affected by UV light and temperatures over 121 °C, so consumers should not fully rely on the inks message when it comes to deciding the proper time to consume a food.

Packaging Machines

Depending on the type of food being packed, packing comes in various types. To pack these food materials, various food packaging machines are used. The packing styles also change depending on the storage life of the product.

Food that are high perishable like fresh processed meats and frozen items are best when vacuum packed since it can tremendously extend its storage life. There is a separate type of food packaging machine or food packing equipment used to perform vacuum packaging of the products.

Here are the various types of food packaging machines:

Food Vacuum Packaging Machine

It is one of the most efficient packaging machine to pack foods because it avoids air making food remain fresh. As aerobic microorganisms are responsible in swift deterioration of foods, they hardly thrive or are immobilized under this condition.

Food vacuum packaging machine helps to extend storage life of food products thereby making the product well suited for sale on the freezer or cold display storage units of several retail stores.

Biscuit Packaging Machine

Biscuit packaging machine is another type of food packaging equipment. It is usually fitted with electronic digital temperature controller to maintain high precision in achieving the desired temperature during food packing process.

It helps to bring optimum freshness of the food. The most interesting aspect of this machine is that packaged products are closely monitored with its automatic feed counter that shows the quantity of items placed packed by machine. This makes it easier for the food manufacturing companies to monitor daily factory output.

Bundling Food Packaging Equipment

Bundling Food packaging Equipment is quite common and is widely used by many food suppliers. It is capable of storing huge quantity of foods before they are banded or wrapped together as a single bundle.

It is also called as the banding machine. It can also be used for packing small items such as stick candies or individually packed hot-dogs that need to be bundled together for economic purposes.

Bagging Machine

It is popular in several China food processing factories. Foods in this case are packed in bags, sacks and pouches. This bagging machine is common to pack cereals and powdered foods such as milk powder and sugar.

Closing Machines

These closing machines are similarly common in many food factories. This equipment is used to tie metal wires to enclose the food bag or pouch.

Capping Machines

Capping machines are popular among food suppliers of food syrups and drinks. This equipment is not used solely to pack food items but it is usually used in conjunction with other food packaging equipment.

The major function of this equipment is to close bottled food items by placing air-tight caps. This is common in soda-manufacturing companies.

Accumulation Machinery

Accumulation machinery is used along with the capping machine. This machine allows proper alignment of bottles for systematic and organized filling of foods. It is used in soda companies and bottled-water companies.

There are various types of food packaging machines. It is important to tailor-fit the selection when yo buy one of these machineries according to the type of food that is being packaged to ensure optimum quality products with fully extended storage life

Symbols used on Packages

Food products can have symbols related to ingredients, nutritional information, dietary restrictions, and allergens. Similarly, cosmetic products can have symbols for certain ingredients and allergens, but also for things like ethical production and testing.

There are so many symbols that exist out in the marketplace, especially if you include international products, but here are some of the most common symbols found on food and cosmetic products here in the U.S.

Food Symbols

Non-GMO Project

The Non-GMO Project seal verifies that the product was made in an environment dedicated to NO *Genetically modified* organisms (*GMOs*). GMOs are living organisms whose genetic material has been manipulated in a laboratory through genetic engineering, creating combinations of plant and animal genes that do not occur in nature. The Non-GMO Project offers independent verification of testing and GMO controls for U.S. and Canada based products.

Certified Gluten-free

This logo, created by the Gluten Free Certification Organization (GFCO), verifies that the food within the package contains no traces of gluten and that the product was not exposed to cross-contamination.

Certified Vegan

This logo has been added to product packaging as a response to the growing number of consumers interested in vegan products. Products with this label are verified to be free from any animal products or byproducts.

USDA Organic

Produced by the USDA's National Organic Program, this label represents the regulations and standards for products that are labeled as being produced organically. The products must have been produced in an environment that promotes sustainability where no synthetic fertilizers, irradiation, and genetic engineering is present.

Fair Trade Certified

This label can be found on honey, tea, chocolate, coffee, nuts, and grains. It certifies that the product was produced sustainably and fairly, benefiting the farmers and workers involved in its production.

Rainforest Alliance Certified

This label, created by The Rainforest Alliance, can be found on coffee, tea, chocolate, and fruit juice, among other products. It ensures that the products were produced sustainably, with the smallest possible impact on the environment.

OK Kosher Certification

The label represents the OK Kosher Certification (the 'O' being the circle). This symbol is used internationally and can be found on thousands of popular products from major companies. Foods with this label have been certified by the organization's 350 kosher experts.

OU Kosher Certification

Also signifying Kosher food, this OU label was created by the Orthodox Union to verify which modern foods abide by ancient dietary laws. This label is used internationally for hundreds of thousands of products and is the strictest kosher verification system. Because this labeling system is so strict, there are different variations of the OU that can be found on products.

These variations include:

OU — The food contains neither meat nor dairy, a "neutral" product.

OU-D — Dairy product.

OU-M — The product is made with meat or meat ingredients.

OU-F — The product is made with fish ingredients.

Heart Check Symbol

Foods with packages marked with this symbol meet the American Heart Association's requirements of low amounts of saturated fat and cholesterol for people over the age of two. Additionally, this symbol supports the American Heart Association's food science and recommendations.

Whole Grain Council Stamp

This stamp is put on Pack by Whole Grains Council. There are two different versions of this stamp, the Basic Stamp and the 100% Stamp. The product will only have the 100% Stamp if 100% of all the grains used in the product are whole grains. If all the grains are not whole grains but the product contains at least eight grams of whole grains, the product is given a Basic Stamp.

Animal Welfare Approved

This label for meat, dairy, and poultry products certifies that the animals were raised humanely and in an environmentally-friendly manner. This independent, non-profit recognizes farms for maintaining the highest animal welfare and environmental standards.

The American Humane Certified Program

Since 1877, American Humane Association has been ensuring the welfare, wellness and well-being of children and animals. They have been dedicated to improving child and animal welfare and have been at the forefront of every major progressive effort to protect children, farm animals, and companion animals from abuse and neglect.

Global Animal Partnership

Whole Foods created the 5-step program called Global Animal Partnership, a certification program it requires its vendors to use. Some small retailers have also begun using.

Certified Humane

This label was created by the Humane Farm Animal Care (HFAC) and, like the one above, it certifies that the animal products came from facilities that were proven to be raising and caring for their animals humanely.

Many also companies create their own standards and create a label for their products to create consumer trust. One example is the brand Happy Egg. They created their own clever symbol, call Happy Certified. They go beyond humane standards to create an ecosystem for their hens to flourish and thrive. They state they are committed to raising hens and eggs with love and make an extra effort to improve their hens' welfare is what makes them happy calling their certification, Happy Certified. Beware, this kind of certification creation, does not always end happy. If it's not back up by true claims, it can backfire like it did for the "Smart Choices" program.

Cosmetic Symbols

PETA Bunny

PETA, an animal-welfare organization, created this symbol for verified companies to indicate that their products are strictly cruelty-free and are not tested on animals during any part of production.

Leaping Bunny

Created by The Leaping Bunny Program, this label certifies that the product is cruelty-free and was not tested on animals. In order to display this label on their products, companies must take a pledge that none of their products or any ingredients have been tested on animals.

Period after Opening (POA) Symbol

Since cosmetics products can degrade over time and can cause products to go bad including causing skin infections, this label indicates the shelf-life of a product after opening before it is considered to be expired and should be thrown away. The number followed by the M stands for the specific number of months the product is good after opening. This information is typically given to you by your manufacturer.

This symbol, which can be shown on any type of products in addition to cosmetics, is normally found with product information on the package or product itself. It communicates that you are only seeing a portion of the total product information and might have to refer to a different part of the package or product for the rest of the information.

Estimated Symbol

This E symbol indicates that the product was filled using "average fill system". So, if this E appears under the amount in grams or milliliters on the package, it means that the product contains the amount advertised. This must appear on all products sold in the EU.

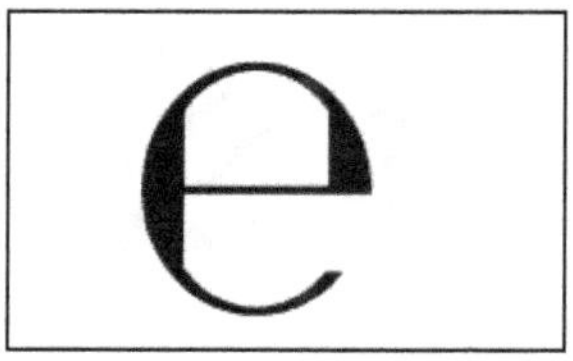

Green Dot

This green label is meant to show that the manufacturer of the product pays to recover and recycle it. While this label is used for recycling internationally, the program is only in Europe. This label is not meant to replace the general recycling symbol, but to be used in addition to it.

Flammable

This symbol on a packaging indicates that the product itself or at least one of its ingredients is flammable if it is exposed to high heat or a flame. It serves as a warning to keep the product away from sources of heat.

Resin Identification

One of these six symbols, created and used by Society of the Plastics Industry (SPI), is usually found somewhere on plastic products. These symbols identify what type of polymer resin the plastic product is made out of so plastics of the same polymer resin types can be recycled properly together.

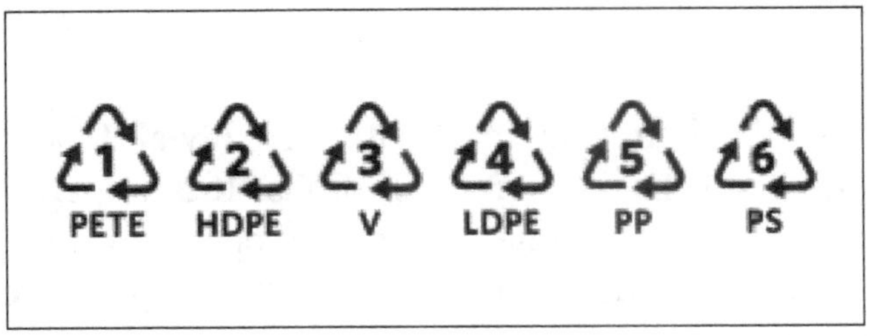

Materials for Food Packaging

Package design and construction play a significant role in determining the shelf life of a food product. The right selection of packaging materials and technologies maintains product quality and freshness during distribution and storage. Materials that have traditionally been used in food packaging include glass, metals (aluminum, foils and laminates, tinplate, and tin-free steel), paper and paperboards, and plastics. Moreover, a wider variety of plastics have been introduced in both rigid and flexible forms. Today's food packages often combine several materials to exploit each material's functional or aesthetic properties. As research to improve food packaging continues, advances in the field may affect the environmental impact of packaging.

The primary method of regulation is through the food contact notification process that requires that manufacturers notify FDA 120 d prior to marketing a food contact substance (FCS) for a new use. An FCS is "any substance intended for use as a component of materials used in manufacturing, packing, packaging, transporting or holding of food if the use is not intended to have a technical effect in such food". All FCSs that may reasonably migrate to food under conditions of intended use are identified and regulated as food additives unless classified as generally recognized as safe (GRAS) substances.

Glass

Glass has an extremely long history in food packaging; the 1st glass objects for holding food are believed to have appeared around 3000 BC. The production of glass containers involves heating a mixture of silica (the glass former), sodium carbonate (the melting agent), and limestone/calcium carbonate and alumina (stabilizers) to high temperatures until the materials melt into a thick liquid mass that is then poured into molds. Recycled broken glass (cullet) is also used in glass manufacture and may account for as much as 60% of all raw materials. Glass containers used in food packaging are often surface-coated to provide lubrication in the production line and eliminate scratching or surface abrasion and line jams. Glass coatings also increase and preserve the strength of the bottle to reduce breakage. Improved break resistance allows manufacturers to use thinner glass, which reduces weight and is better for disposal and transportation.

Because it is odorless and chemically inert with virtually all food products, glass has several advantages for food- packaging applications: It is impermeable to gases and vapors, so it maintains product freshness for a long period of time without impairing taste or flavor. The ability to withstand high processing temperatures makes glass useful for heat sterilization of both low- acid and high-acid foods. Glass is rigid, provides good insulation, and can be produced in numerous different shapes. The transparency of glass allows consumers to see the product, yet variations in glass color can protect light-sensitive contents. Finally, glass packaging benefits the environment because it is reusable and recyclable.

Like any material, glass has some disadvantages. Despite efforts to use thinner glass, its heavy weight adds to transportation costs. Another concern is its brittleness and susceptibility to breakage from internal pressure, impact, or thermal shock.

Metal

Metal is the most versatile of all packaging forms. It offers a combination of excellent physical protection and barrier properties, formability and decorative potential, recyclability, and consumer acceptance. The 2 metals most predominantly used in packaging are aluminum and steel.

Aluminum

Commonly used to make cans, foil, and laminated paper or plastic packaging, aluminum is a lightweight, silvery white metal derived from bauxite ore, where it exists in combination with oxygen as alumina. Magnesium and manganese are often added to aluminum to improve its strength properties. Unlike many metals, aluminum is highly resistant to most forms of corrosion; its natural coating of aluminum oxide provides a highly effective barrier to the effects of air, temperature, moisture, and chemical attack.

Besides providing an excellent barrier to moisture, air, odors, light, and microorganisms, aluminum has good flexibility and surface resilience, excellent malleability and formability, and outstanding embossing potential. It is also an ideal material for recycling because it is easy to reclaim and process into new products. Pure aluminum is used for light packaging of primarily soft-drink cans, pet food, seafood, and prethreaded closures. The main disadvantages of aluminum are its high cost compared to other metals (for example, steel) and its inability to be welded, which renders it useful only for making seamless containers.

Aluminum Foil

Aluminum foil is made by rolling pure aluminum metal into very thin sheets, followed by annealing to achieve dead-folding properties (a crease or fold made in the film will stay in place), which allows it to be folded tightly. Moreover, aluminum foil is available in a wide range of thicknesses, with thinner foils used to wrap food and thicker foils used for trays. Like all aluminum packaging, foil provides an excellent barrier to moisture, air, odors, light, and microorganisms. It is inert to acidic foods and does not require lacquer or other protection. Although aluminum is easily recyclable, foils cannot be made from recycled aluminum without pinhole formation in the thin sheets.

Laminates and Metallized Films

Lamination of packaging involves the binding of aluminum foil to paper or plastic film to improve barrier properties. Thin gauges facilitate application. Although lamination to plastic enables heat sealability, the seal does not completely bar moisture and air. Because laminated aluminum is relatively expensive, it is typically used to package high value foods such as dried soups, herbs, and spices. A less expensive alternative to laminated packaging is metallized film. Metallized films are plastics containing a thin layer of aluminum metal. These films have improved barrier properties to moisture, oils, air, and odors, and the highly reflective surface of the aluminum is attractive to consumers. More flexible than laminated films, metallized films are mainly used to package snacks. Although the individual components of laminates and metallized films are technically recyclable, the difficulty in sorting and separating the material precludes economically feasible recycling.

Tinplate

Produced from low-carbon steel (that is, blackplate), tinplate is the result of coating both sides of blackplate with thin layers of tin. The coating is achieved by dipping sheets of steel in molten tin (hot-dipped tinplate) or by the electro-deposition of tin on the steel sheet (electrolytic tinplate). Although tin provides steel with some corrosion resistance, tinplate containers are often lacquered to provide an inert barrier between the metal and the food product. Commonly used lacquers are materials in the epoxy phenolic and oleoresinous groups and vinyl resins.

In addition to its excellent barrier properties to gases, water vapor, light, and odors, tinplate can be heat-treated and sealed hermetically, making it suitable for sterile products. Because it has good ductility and formability, tinplate can be used for containers of many different shapes. Thus, tinplate is widely used to form cans for drinks, processed foods, and aerosols; containers for powdered foods and sugar-or flour-based confections; and as package closures. Tinplate is an excellent substrate for modern metal coating and lithoprinting technology, enabling outstanding graphical decoration. Its relatively low weight and high mechanical strength make it easy to ship and store. Finally, tinplate is easily recycled many times without loss of quality and is significantly lower in cost than aluminum.

Tin-free Steel

Also known as electrolytic chromium or chrome oxide coated steel, tin-free steel requires a coating of organic material to provide complete corrosion resistance. Even though the chrome/chrome oxide makes tin-free steel unsuitable for welding, this property makes it excellent for adhesion of coatings such as paints, lacquers, and inks. Like tinplate, tin-free steel has good formability and strength, but it is marginally less expensive than tinplate. Food cans, can ends, trays, bottle caps, and closures can all be made from tin-free steel. In addition, it can also be used to make large containers (such as drums) for bulk sale and bulk storage of ingredients or finished goods.

Plastics

Plastics are made by condensation polymerization (polycondensation) or addition polymerization (polyaddition) of monomer units. In polycondensation, the polymer chain grows by condensation reactions between molecules and is accompanied by formation of low molecular weight byproducts such as water and methanol. Polycondensation involves monomers with at least 2 functional groups such as alcohol, amine, or carboxylic groups. In polyaddition, polymer chains grow by addition reactions, in which 2 or more molecules combine to form a larger molecule without liberation of by-products. Polyaddition involves unsaturated monomers; double or triple bonds are broken to link monomer chains. There are several advantages to using plastics for food packaging. Fluid and moldable, plastics can be made into sheets, shapes, and structures, offering considerable design flexibility. Because they are chemically resistant, plastics are inexpensive and lightweight with a wide range of physical and optical properties. In fact, many plastics are heat sealable, easy to print, and can be integrated into production processes where the package is formed, filled, and sealed in the same production line.

The major disadvantage of plastics is their variable permeability to light, gases, vapors, and low molecular weight molecules.

There are 2 major categories of plastics: Thermosets and thermoplastics. Thermosets are polymers that solidify or set irreversibly when heated and cannot be remolded. Because they are strong and durable, they tend to be used primarily in automobiles and construction applications such as adhesives and coatings, not in food packaging applications. On the other hand, thermoplastics are polymers that soften upon exposure to heat and return to their original condition at room temperature. Because thermoplastics can easily be shaped and molded into various products such as bottles, jugs, and plastic films, they are ideal for food packaging. Moreover, virtually all thermoplastics are recyclable (melted and reused as raw materials for production of new products), although separation poses some practical limitations for certain products. The recycling process requires separation by resin type as identified by the American Plastics Council.

Table: Resin identification codes for plastic recycling.

Resin	Code	Amount generated (thousand tons)	Amount recycled (thousand tons)
Polyethylene terephthalate	1	2860	540
High-density polyethylene	2	5890	520
Polyvinyl chloride	3	1640	
Low-density polyethylene	4	6450	190[a]
Polypropylene	5	4000	10
Polystyrene	6	2590	
Other resins	7	5480	390

[a]Includes linear low-density polyethylene.

There have been some health concerns regarding residual monomer and components in plastics, including stabilizers, plasticizers, and condensation components such as bisphenol A. Some of these concerns are based on studies using very high intake levels; others have no scientific basis. To ensure public safety, FDA carefully reviews and regulates substances used to make plastics and other packaging materials. Any substance that can reasonably be expected to migrate into food is classified as an indirect food additive subject to FDA regulations. A threshold of regulation—defined as a specific level of dietary exposure that typically induces toxic effects and therefore poses negligible safety concerns—may be used to exempt substances used in food contact materials from regulation as food additives. FDA revisits the threshold level if new scientific information raises concerns. Furthermore, FDA advises consumers to use plastics for intended purposes in accordance with the manufacturer's directions to avoid unintentional safety concerns.

Despite these safety concerns, the use of plastics in food packaging has continued to increase due to the low cost of materials and functional advantages (such as thermosealability, microwavability, optical properties, and unlimited sizes and shapes) over traditional materials such as glass and tinplate. Multiple types of plastics are being used as materials for packaging food, including polyolefin, polyester, polyvinyl chloride, polyvinylidene chloride, polystyrene, polyamide, and ethylene vinyl alcohol. Although more than 30 types of plastics have been used as packaging materials, polyolefins and polyesters are the most common.

Polyolefins

Polyolefin is a collective term for polyethylene and polypropylene, the 2 most widely used plastics in food packaging, and other less popular olefin polymers. Polyethylene and polypropylene both possess a successful combination of properties, including flexibility, strength, lightness, stability, moisture and chemical resistance, and easy processability, and are well suited for recycling and reuse.

The simplest and most inexpensive plastic made by addition polymerization of ethylene is polyethylene. There are 2 basic categories of polyethylene: high density and low density. High-density polyethylene is stiff, strong, tough, resistant to chemicals and moisture, permeable to gas, easy to process, and easy to form. It is used to make bottles for milk, juice, and water; cereal box liners; margarine tubs; and grocery, trash, and retail bags. Low-density polyethylene is flexible, strong, tough, easy to seal, and resistant to moisture. Because low-density polyethylene is relatively transparent, it is predominately used in film applications and in applications where heat sealing is necessary. Bread and frozen food bags, flexible lids, and squeezable food bottles are examples of low-density polyethylene. Polyethylene bags are sometimes reused (both for grocery and nongrocery retail). Of the 2 categories of polyethylene, high-density polyethylene containers, especially milk bottles, are the most recycled among plastic packages.

Harder, denser, and more transparent than polyethylene, polypropylene has good resistance to chemicals and is effective at barring water vapor. Its high melting point (160 °C) makes it suitable for applications where thermal resistance is required, such as hot-filled and microwavable packaging. Popular uses include yogurt containers and margarine tubs. When used in combination with an oxygen barrier such as ethylene vinyl alcohol or polyvinylidene chloride, polypropylene provides the strength and moisture barrier for catsup and salad dressing bottles.

Polyesters

Polyethylene terephthalate (PET or PETE), polycarbonate, and polyethylene naphthalate (PEN) are polyesters, which are condensation polymers formed from ester monomers that result from the reaction between carboxylic acid and alcohol. The most commonly used polyester in food packaging is PETE.

Polyethylene Terephthalate

Formed when terephthalic acid reacts with ethylene glycol, PETE provides a good barrier to gases (oxygen and carbon dioxide) and moisture. It also has good resistance to heat, mineral oils, solvents, and acids, but not to bases. Consequently, PETE is becoming the packaging material of choice for many food products, particularly beverages and mineral waters. The use of PETE to make plastic bottles for carbonated drinks is increasing steadily. The main reasons for its popularity are its glass-like transparency, adequate gas barrier for retention of carbonation, light weight, and shatter resistance. The 3 major packaging applications of PETE are containers (bottles, jars, and tubs), semirigid sheets for thermoforming (trays and blisters), and thin-oriented films (bags and snack food wrappers). PETE exists both as an amorphous (transparent) and a semicrystalline (opaque and white) thermoplastic material. Amorphous PETE has better ductility but less stiffness and hardness than semicrystalline PETE, which has good strength, ductility, stiffness, and hardness. Recycled PETE from soda bottles is used as fibers, insulation, and other nonfood packaging applications.

Polycarbonate

Polycarbonate is formed by polymerization of a sodium salt of bisphenol acid with carbonyl dichloride (phosgene). Clear, heat resistant, and durable, it is mainly used as a replacement for glass in items such as large returnable/refillable water bottles and sterilizable baby bottles. Care must be taken when cleaning polycarbonate because using harsh detergents such as sodium hypochlorite is not recommended because they catalyze the release of bisphenol A, a potential health hazard. An extensive literature analysis by suggests the need for a new risk assessment for the low-dose effects of this compound.

Polyethylene Naphthalate

PEN is a condensation polymer of dimethyl naphthalene dicarboxylate and ethylene glycol. It is a relatively new member of the polyester family with excellent performance because of its high glass transition temperature. PEN's barrier properties for carbon dioxide, oxygen, and water vapor are superior to those of PETE, and PEN provides better performance at high temperatures, allowing hot refills, rewashing, and reuse. However, PEN costs 3 to 4 times more than PETE. Because PEN provides protection against transfer of flavors and odors, it is well suited for manufacturing bottles for beverages such as beer.

Polyvinyl Chloride

Polyvinyl chloride (PVC), an addition polymer of vinyl chloride, is heavy, stiff, ductile, and a medium strong, amorphous, transparent material. It has excellent resistance to chemicals (acids and bases), grease, and oil; good flow characteristics; and stable electrical properties. Although PVC is primarily used in medical and other nonfood applications, its food uses include bottles and packaging films. Because it is easily thermoformed, PVC sheets are widely used for blister packs such as those for meat products and unit dose pharmaceutical packaging.

PVC can be transformed into materials with a wide range of flexibility with the addition of plasticizers such as phthalates, adipates, citrates, and phosphates. Phthalates are mainly used in nonfood packaging applications such as cosmetics, toys, and medical devices. Safety concerns have emerged over the use of phthalates in certain products, such as toys. Because of these safety concerns, phthalates are not used in food packaging materials in the United States; instead, alternative nonphthalate plasticizers such as adipates are used. For example, di-(2-ethylhexyl) adipate (DEHA) is used in the manufacture of plastic cling wraps. These alternative plasticizers also have the potential to leach into food but at lower levels than phthalates. Low levels of DEHA have shown no toxicity in animals. Finally, PVC is difficult to recycle because it is used for such a variety of products, which makes it difficult to identify and separate. In addition, incineration of PVC presents environmental problems because of its chlorine content.

Polyvinylidene Chloride

Polyvinylidene chloride (PVdC) is an addition polymer of vinylidene chloride. It is heat sealable and serves as an excellent barrier to water vapor, gases, and fatty and oily products. It is used in flexible packaging as a monolayer film, a coating, or part of a co-extruded product. Major applications include packaging of poultry, cured meats, cheese, snack foods, tea, coffee, and confectionary. It is also used in hot filling, retorting, low-temperature storage, and modified atmosphere packaging. PVdC contains twice the amount of chlorine as PVC and therefore also presents problems with incineration.

Polystyrene

Polystyrene, an addition polymer of styrene, is clear, hard, and brittle with a relatively low melting point. It can be mono-extruded, co-extruded with other plastics, injection molded, or foamed to produce a range of products. Foaming produces an opaque, rigid, lightweight material with impact protection and thermal insulation properties. Typical applications include protective packaging such as egg cartons, containers, disposable plastic silverware, lids, cups, plates, bottles, and food trays. In expanded form, polystyrene is used for nonfood packaging and cushioning, and it can be recycled or incinerated.

Polyamide

Commonly known as nylon (a brand name for a range of products produced by DuPont), polyamides were originally used in textiles. Formed by a condensation reaction between diamine and diacid, polyamides are polymers in which the repeating units are held together by amide links. Different types of polyamides are characterized by a number that relates to the number of carbons in the originating monomer. For example, nylon-6 has 6 carbons and is typically used in packaging. It has mechanical and thermal properties similar to PETE, so it has similar usefulness, such as boil-in bag packaging. Nylon also offers good chemical resistance, toughness, and low gas permeability.

Ethylene Vinyl Alcohol

Ethylene vinyl alcohol (EVOH) is a copolymer of ethylene and vinyl alcohol. It is an excellent barrier to oil, fat, and oxygen. However, EVOH is moisture sensitive and is thus mostly used in multilayered co-extruded films in situation where it is not in direct contact with liquids.

Laminates and Co-extrusions

Plastic materials can be manufactured either as a single film or as a combination of more than 1 plastic. There are 2 ways of combining plastics: lamination and co-extrusion. Lamination involves bonding together 2 or more plastics or bonding plastic to another material such as paper or aluminum. Bonding is commonly achieved by use of water-, solvent-, or solids-based adhesives. After the adhesives are applied to 1 film, 2 films are passed between rollers to pressure bond them together. Lamination using laser rather than adhesives has also been used for thermoplastics. Lamination enables reverse printing, in which the printing is buried between layers and thus not subject to abrasion, and can add or enhance heat sealability.

In co-extrusion, 2 or more layers of molten plastics are combined during the film manufacture. This process is more rapid (requires 1 step in comparison to multiple steps with lamination) but requires materials that have thermal characteristics that allow co-extrusion. Because co-extrusion and lamination combine multiple materials, recycling is complicated. However, combining materials results in the additive advantage of properties from each individual material and often reduces the total amount of packaging material required. Therefore, co-extrusion and lamination can be sources of packaging reduction.

Paper and Paperboard

The use of paper and paperboards for food packaging dates back to the 17th century with accelerated usage in the later part of the 19th century. Paper and paperboard are sheet materials made from an interlaced network of cellulose fibers derived from wood by using sulfate and sulfite. The fibers are then pulped and bleached and treated with chemicals such as slimicides and strengthening agents to produce the paper product. FDA regulates the additives used in paper and paperboard food packaging (21 CFR Part 176). Paper and paperboards are commonly used in corrugated boxes, milk cartons, folding cartons, bags and sacks, and wrapping paper. Tissue paper, paper plates, and cups are other examples of paper and paperboard products.

Paper

Plain paper is not used to protect foods for long periods of time because it has poor barrier properties and is not heat sealable. When used as primary packaging (that is, in contact with food), paper is almost always treated, coated, laminated, or impregnated with materials such as waxes, resins, or lacquers to improve functional and protective properties. The many different types of paper used in food packaging are as follows:

- Kraft paper—Produced by a sulfate treatment process, kraft paper is available in several forms: natural brown, unbleached, heavy duty, and bleached white. The natural kraft is the strongest of all paper and is commonly used for bags and wrapping. It is also used to package flour, sugar, and dried fruits and vegetables.
- Sulfite paper—Lighter and weaker than kraft paper, sulfite paper is glazed to improve its appearance and to increase its wet strength and oil resistance. It can be coated for higher print quality and is also used in laminates with plastic or foil. It is used to make small bags or wrappers for packaging biscuits and confectionary.
- Greaseproof paper—Greaseproof paper is made through a process known as beating, in which the cellulose fibers undergo a longer than normal hydration period that causes the fibers to break up and become gelatinous. These fine fibers then pack densely to provide a surface that is resistant to oils but not wet agents. Greaseproof paper is used to wrap snack foods, cookies, candy bars, and other oily foods, a use that is being replaced by plastic films.
- Glassine—Glassine is greaseproof paper taken to an extreme (further hydration) to produce a very dense sheet with a highly smooth and glossy finish. It is used as a liner for biscuits, cooking fats, fast foods, and baked goods.
- Parchment paper—Parchment paper is made from acid-treated pulp (passed through a sulfuric acid bath). The acid modifies the cellulose to make it smoother and impervious to water and oil, which adds some wet strength. It does not provide a good barrier to air and moisture, is not heat sealable, and is used to package fats such as butter and lard.

Paperboard

Paperboard is thicker than paper with a higher weight per unit area and often made in multiple layers. It is commonly used to make containers for shipping—such as boxes, cartons, and trays—and seldom used for direct food contact. The various types of paperboard are as follows:

- White board—Made from several thin layers of bleached chemical pulp, white board is typically used as the inner layer of a carton. White board may be coated with wax or laminated with polyethylene for heat sealability, and it is the only form of paperboard recommended for direct food contact.

- Solid board—Possessing strength and durability, solid board has multiple layers of bleached sulfate board. When laminated with polyethylene, it is used to create liquid cartons (known as milk board). Solid board is also used to package fruit juices and soft drinks.
- Chipboard—Chipboard is made from recycled paper and often contains blemishes and impurities from the original paper, which makes it unsuitable for direct contact with food, printing, and folding. It is often lined with white board to improve both appearance and strength. The least expensive form of paperboard, chipboard is used to make the outer layers of cartons for foods such as tea and cereals.
- Fiberboard—Fiberboard can be solid or corrugated. The solid type has an inner white board layer and outer kraft layer and provides good protection against impact and compression. When laminated with plastics or aluminum, solid fiberboard can improve barrier properties and is used to package dry products such as coffee and milk powder. The corrugated type, also known as corrugated board, is made with 2 layers of kraft paper with a central corrugating (or fluting) material. Fiberboard's resistance to impact abrasion and crushing damage makes it widely used for shipping bulk food and case packing of retail food products.

Paper Laminates

Paper laminates are coated or uncoated papers based on kraft and sulfite pulp. They can be laminated with plastic or aluminum to improve various properties. For example, paper can be laminated with polyethylene to make it heat sealable and to improve gas and moisture barrier properties. However, lamination substantially increases the cost of paper. Laminated paper is used to package dried products such as soups, herbs, and spices.

5

Food Allergens, Contamination and Illness

Allergens are the naturally-occurring proteins in food which cause irrational immune responses. The presence of some unwanted chemicals and harmful microorganisms in the food causes food contamination, food-borne illnesses and food allergies. The topics elaborated in this chapter will help in gaining a better perspective about food allergens, contamination and illness.

Food Allergens

Food allergens are typically naturally-occurring proteins in foods or derivatives of them that cause abnormal immune responses. Prevalence of food allergies around the world is believed to be increasing, with more than 8% of children and 2% of adults in countries like Australia and New Zealand having allergy to one or more foods. The most common allergens for young children are milk and egg but, fortunately, many children outgrow these allergies by the time they have reached 5-7 years of age. On the other hand, allergies such as those to seafood, peanut and tree nut may develop later and are lifelong conditions.

Practically all foods have the capacity to cause an allergic reaction in a person who has become sensitised to proteins in it. However, in Australia and New Zealand there are 9 foods or food groups that cause about 90% of all allergic reactions: Peanuts; tree nuts; soy; milk, egg; cereals; seafood; fish; and sesame.

Allergic reactions to foods vary greatly from mild gastrointestinal discomfort, to skin rashes and potentially life threatening asthma and anaphylaxis. Commonly many adverse reactions to food are referred to collectively as food allergies. However, true food allergies represent only a fraction of the diverse range of individualistic adverse reactions to foods, that also include food intolerances.

Some consumers may also experience mild allergic symptoms to fresh fruits and vegetables such as kiwi, apples, peaches, melons, pineapple and papaya. This condition, known as oral allergy syndrome or pollen-food syndrome, is normally associated with a primary allergy to pollen (e.g. birch, ragwort or grasses) or latex. In these individuals, the immune system reacts to the food proteins as if they were pollen and the symptoms are generally limited to the mouth and throat.

There is currently no cure for food allergies but effective care and emergency treatments are available. The only successful method to manage a food allergy is avoidance of all foods containing the allergen.

Living with an allergy, either for you or someone in your family, requires great compromise to the quality of life. It takes longer to find products in the supermarket that are safe to eat because of the need to study food labels and scrutinise ingredient lists, and it costs more because generally cheaper products and house branded foods have 'may contain'-like allergen statements on the label.

Despite food allergies affecting only a small proportion of the population, risk management and mandatory product labelling for the key food allergens are critical food safety matters for businesses in the food industry. This is where a responsible food industry plays an absolutely critical role.

Symptoms

For some people, an allergic reaction to a particular food may be uncomfortable but not severe. For other people, an allergic food reaction can be frightening and even life-threatening. Food allergy symptoms usually develop within a few minutes to two hours after eating the offending food.

The most common food allergy signs and symptoms include:

- Tingling or itching in the mouth.
- Hives, itching or eczema.
- Swelling of the lips, face, tongue and throat or other parts of the body.
- Wheezing, nasal congestion or trouble breathing.
- Abdominal pain, diarrhea, nausea or vomiting.
- Dizziness, lightheadedness or fainting.

Anaphylaxis

In some people, a food allergy can trigger a severe allergic reaction called anaphylaxis. This can cause life-threatening signs and symptoms, including:

- Constriction and tightening of the airways.
- A swollen throat or the sensation of a lump in your throat that makes it difficult to breathe.
- Shock with a severe drop in blood pressure.
- Rapid pulse.
- Dizziness, lightheadedness or loss of consciousness.

Emergency treatment is critical for anaphylaxis. Untreated, anaphylaxis can cause a coma or even death.

See a doctor or allergist if you have food allergy symptoms shortly after eating. If possible, see your doctor when the allergic reaction is occurring. This will help your doctor make a diagnosis.

Seek emergency treatment if you develop any signs or symptoms of anaphylaxis, such as:

- Constriction of airways that makes it difficult to breathe.
- Shock with a severe drop in blood pressure.
- Rapid pulse.
- Dizziness or lightheadedness.

Causes

When you have a food allergy, your immune system mistakenly identifies a specific food or a substance in food as something harmful. In respose, your immune system triggers cells to release an antibody known as immunoglobulin E (IgE) to neutralize the allergy-causing food or food substance (the allergen).

The next time you eat even the smallest amount of that food, IgE antibodies sense it and signal your immune system to release a chemical called histamine, as well as other chemicals, into your bloodstream. These chemicals cause allergy symptoms.

In adults, the majority of food allergies are triggered by certain proteins in:

- Shellfish, such as shrimp, lobster and crab.
- Peanuts.
- Tree nuts, such as walnuts and pecans.
- Fish.

In children, food allergies are commonly triggered by proteins in:

- Peanuts.
- Tree nuts.
- Eggs.
- Cow's milk.
- Wheat.
- Soy.

Pollen-food Allergy Syndrome

Also known as oral allergy syndrome, pollen-food allergy syndrome affects many people who have hay fever. In this condition, certain fresh fruits and vegetables or nuts and spices can trigger an allergic reaction that causes the mouth to tingle or itch. In serious cases, the reaction results in swelling of the throat or even anaphylaxis.

Proteins in certain fruits, vegetables, nuts and spices cause the reaction because they're similar to allergy-causing proteins found in certain pollens. This is an example of cross-reactivity.

When you cook foods that trigger pollen-food allergy syndrome, your symptoms may be less severe.

This following table shows the specific fruits, vegetables, nuts and spices that can cause pollen-food allergy syndrome in people who are allergic to different pollens.

If you are allergic to:	Birch pollen	Ragweed pollen	Grasses	Mugwort pollen
You may also have a reaction to:	Almond Apple Apricot Carrot Celery Cherry Hazelnut Peach Peanut Pear Plum Raw potatoes Soybean Some herbs and spices (anise, caraway, coriander, fennel, parsley)	Bananas Cucumber Melons (cantaloupe, honeydew and watermelon) Zucchini	Cucumber Kiwi Melons (cantaloupe, honeydew and watermelon) Orange Peanut Tomatoes White potato Zucchini	Apples Bell pepper Broccoli Cabbage Carrots Celery Cauliflower Garlic Onion Peach Some herbs and spices (anise, black pepper, caraway seed, coriander, fennel, mustard, parsley)

Exercise-induced Food Allergy

Eating certain foods may cause some people to feel itchy and lightheaded soon after starting to exercise. Serious cases may even involve hives or anaphylaxis. Not eating for a couple of hours before exercising and avoiding certain foods may help prevent this problem.

Food Intolerance and other Reactions

A food intolerance or a reaction to another substance you ate may cause the same signs and symptoms as a food allergy does — such as nausea, vomiting, cramping and diarrhea.

Depending on the type of food intolerance you have, you may be able to eat small amounts of problem foods without a reaction. By contrast, if you have a true food allergy, even a tiny amount of food may trigger an allergic reaction.

One of the tricky aspects of diagnosing food intolerance is that some people are sensitive not to the food itself but to a substance or ingredient used in the preparation of the food.

Common conditions that can cause symptoms mistaken for a food allergy include:

- Absence of an enzyme needed to fully digest a food: You may not have adequate amounts of some enzymes needed to digest certain foods. Insufficient quantities of the enzyme lactase, for example, reduce your ability to digest lactose, the main sugar in milk products. Lactose intolerance can cause bloating, cramping, diarrhea and excess gas.
- Food poisoning: Sometimes food poisoning can mimic an allergic reaction. Bacteria in spoiled tuna and other fish also can make a toxin that triggers harmful reactions.
- Sensitivity to food additives: Some people have digestive reactions and other symptoms after eating certain food additives. For example, sulfites used to preserve dried fruit, canned goods and wine can trigger asthma attacks in sensitive people.
- Histamine toxicity: Certain fish, such as tuna or mackerel, that are not refrigerated properly and that contain high amounts of bacteria may also contain high levels of histamine that trigger symptoms similar to those of food allergy. Rather than an allergic reaction, this is known as histamine toxicity or scombroid poisoning.
- Celiac disease: While celiac disease is sometimes referred to as a gluten allergy, it does not result in anaphylaxis. Like a food allergy, it does involve an immune system response, but it's a unique reaction that's more complex than a simple food allergy.

- This chronic digestive condition is triggered by eating gluten, a protein found in bread, pasta, cookies, and many other foods containing wheat, barley or rye.

If you have celiac disease and eat foods containing gluten, an immune reaction occurs that causes damage to the surface of your small intestine, leading to an inability to absorb certain nutrients.

Risk Factors

Food allergy risk factors include:

- Family history: You're at increased risk of food allergies if asthma, eczema, hives or allergies such as hay fever are common in your family.
- Other allergies: If you're already allergic to one food, you may be at increased risk of becoming allergic to another. Similarly, if you have other types of allergic reactions, such as hay fever or eczema, your risk of having a food allergy is greater.
- Age: Food allergies are more common in children, especially toddlers and infants. As you grow older, your digestive system matures and your body is less likely to absorb food or food components that trigger allergies.

Fortunately, children typically outgrow allergies to milk, soy, wheat and eggs. Severe allergies and allergies to nuts and shellfish are more likely to be lifelong.

Asthma: Asthma and food allergy commonly occur together. When they do, both food allergy and asthma symptoms are more likely to be severe.

Factors that may increase your risk of developing an anaphylactic reaction include:

- Having a history of asthma.
- Being a teenager or younger.
- Delaying use of epinephrine to treat your food allergy symptoms.
- Not having hives or other skin symptoms.

Complications

Complications of food allergy can include:

- Anaphylaxis: This is a life-threatening allergic reaction.
- Atopic dermatitis (eczema): Food allergy may cause a skin reaction, such as eczema.

Prevention

Early introduction of peanut products has been associated with a lower risk of peanut allergy. Before introducing allergenic foods, talk with your child's doctor about the best time to offer them.

However, once food allergy has already developed, the best way to prevent an allergic reaction is to know and avoid foods that cause signs and symptoms. For some people, this is a mere inconvenience, but others find it a greater hardship. Also, some foods — when used as ingredients in certain dishes — may be well-hidden. This is especially true in restaurants and in other social settings.

If you know you have a food allergy, follow these steps:

- Know what you're eating and drinking. Be sure to read food labels carefully.
- If you have already had a severe reaction, wear a medical alert bracelet or necklace that lets others know that you have a food allergy in case you have a reaction and you're unable to communicate.
- Talk with your doctor about prescribing emergency epinephrine. You may need to carry an epinephrine autoinjector (Adrenaclick, EpiPen) if you're at risk of a severe allergic reaction.
- Be careful at restaurants. Be certain your server or chef is aware that you absolutely can't eat the food you're allergic to, and you need to be completely certain that the meal you order doesn't contain it. Also, make sure food isn't prepared on surfaces or in pans that contained any of the food you're allergic to.

Don't be reluctant to make your needs known. Restaurant staff members are usually more than happy to help when they clearly understand your request.

- Plan meals and snacks before leaving home. If necessary, take a cooler packed with allergen-free foods when you travel or go to an event. If you or your child can't have the cake or dessert at a party, bring an approved special treat so no one feels left out of the celebration.

If your child has a food allergy, take these precautions to ensure his or her safety:

- Notify key people that your child has a food allergy.Talk with child care providers, school personnel, parents of your child's friends and other adults who regularly interact with your child. Emphasize that an allergic reaction can be life-threatening and requires immediate action. Make sure that your child also knows to ask for help right away if he or she reacts to food.
- Explain food allergy symptoms. Teach the adults who spend time with your child how to recognize signs and symptoms of an allergic reaction.

- Write an action plan. Your plan should describe how to care for your child when he or she has an allergic reaction to food. Provide a copy of the plan to your child's school nurse and others who care for and supervise your child.
- Have your child wear a medical alert bracelet or necklace. This alert lists your child's allergy symptoms and explains how others can provide first aid in an emergency.

Food Contamination

Food contamination refers to foods that are spoiled or tainted because they either contain microorganisms, such as bacteria or parasites, or toxic substances that make them unfit for consumption.

Food contamination is a serious issue because it results in foodborne diseases that each year affect an estimated seventy-six million people in the United States, while leading to some 325,000 hospitalizations and 5,000 deaths. Hence, awareness of potential sources of food contamination is an important component of good nutrition.

Food contamination can be microbial or environmental, with the former being more common. Environmental contaminants that can enter the food supply chain include pesticides, heavy metals, and other chemical agents. Many opportunities exist for food to become contaminated as it is produced and distributed, bacteria are present in the animals raised for food. Meat and poultry can become contaminated during slaughter through cross-contamination from intestinal fecal matter. Similarly, fresh fruits and vegetables can be contaminated if they are washed using water contaminated with animal manure or human sewage. During food processing, contamination is also possible from infected food handlers. Lastly, poor hygiene in the home is also a factor.

Bacterial Food Contamination

Many bacteria can contaminate food. The most common include the following:

- Campylobacter jejuni: Mishandling of raw poultry and consumption of undercooked poultry are the main causes of C. jejuni contamination.
- Clostridium botulinum: Bacteria producing a toxin in food responsible for botulism, the deadly paralytic nerve illness.
- Escherichia coli: A leading cause of food contamination. Based on a 1999 estimate, 73,000 cases of infection and 61 deaths occur in the United States each year. The E. colio157:H7 strain is found in ground beef, raw milk, chicken, vegetables, and fruit.

- Salmonella typhimurium: Salmonella contamination can occur in meats, poultry, eggs or milk products.
- Shigella: The most common food that these bacteria can contaminate include: salads (potato, chicken, seafood, vegetable), raw vegetables, milk and other dairy products, and meat products especially poultry.
- Staphylococcus aureus: Can be found in custard or cream-filled baked goods, ham, poultry, eggs, potato salad, cream sauces, sandwich fillings.
- Vibrio cholera: These bacteria cause the well-known disease cholera that has caused many outbreaks all over the world. It can be transmitted by water or food.
- Vibrio vulnificus: Free-living ocean bacteria that can cause food borne illnesses from contaminated seafood. Especially dangerous in the warm weather months when eating shellfish that are undercooked or raw.

Spoiled milk is also mostly caused by bacteria such as *Lactococcus cremoris* or *Enterobacter aero-genes,* that cause the milk to form long white strands.

Water contamination is usually due to the presence of three bacteria, *E. coli, Clostridium perfringens,*and enterococci, the bacteria normally found in the feces of people and many animals.

Parasitic Food Contamination

Parasites are organisms that lives in or on a host, and obtain nourishment without benefiting or killing the host. They enter the body through the mouth when contaminated food or drink is swallowed. There are many different types and range in size from single-celled, microscopic organisms (protozoa) to larger, multi-cellular worms (helminths) that can be seen without a microscope. Parasites that contaminate food include:

- *Entamoeba histolytica*: Parasite that causes amoebic dysentery, characterized by severe diarrhea. It is transmitted by contaminated water, and is often called “traveler’s dysentery” because of its prevalence in developing nations.
- *Giardia duodenalis*: Microscopic parasite that can live in the intestines of animals and people. It is found in every region of the world and is one of the most common causes of waterborne and foodborne) illness.
- *Cryptosporidium parvum*: Microscopic parasite, a significant cause of water contamination worldwide. It is found in the intestines of many herd animals including cows, sheep, goats, deer, and elk.
- *Cyclospora cayetanensis*: Single-celled, microscopic parasite. Little is known

about this organism, although cases of infection are being reported from various countries with increasing frequency.

- *Toxoplasma gondii*: Single-celled, microscopic parasite found throughout the world. Found in foods such as raw or undercooked meats, especially pork, lamb, or wild game, and in drinking untreated water.
- *Trichinella spiralis*: Intestinal roundworm whose larvae may migrate from the digestive tract and form cysts in various muscles of the body. In the United States, infections are most prevalent where pork or wild game is consumed raw or undercooked.
- *Taenia saginatajsolium*: *Taenia saginata* (beef tapeworm) and *Taenia solium* (pork tapeworm) are parasitic worms (helminths). They are now uncommon in the United States, although travelers and immigrants are occasionally infected.

Precautions

Simple precautions can reduce the risk of contamination. For instance, meat left at room temperature promotes bacterial growth and refrigeration helps to suppress it. One can also be careful about eating certain foods. Eating raw meats and fish should be avoided as well as salads prepared in restaurants where meats and vegetables share a common surface during preparation.

The Mayo Clinic offers the following advice to prevent food contamination at home:

- Wash hands, utensils and food surfaces often. Keeping hands, utensils and food preparation surfaces clean can prevent cross-contamination, i.e. the transfer of harmful bacteria from one surface to another.
- Keep raw foods separate from ready-to-eat foods. When shopping, preparing food or storing food, keep raw meat, poultry, fish and shellfish away from other foods. This also prevents cross-contamination.
- Cook foods to a safe temperature. You can kill harmful organisms in most foods by cooking them to temperatures between 140 °F and 180 °F.
- Refrigerate or freeze perishable foods. Harmful bacteria can reproduce rapidly if foods are not properly cooled. Refrigerate or freeze perishable foods within two hours of purchasing or preparing them.
- Defrost food safely. Bacteria can reproduce rapidly on meat, poultry and fish at room temperature. To defrost food safely, tightly wrap meat, poultry and fish so that the juices do not drip on other food as they thaw in the refrigerator. Another method is to put the frozen food in a plastic bag and immerse it in cold

water, changing the water every 30 minutes. The sealed food package can also be placed under cold, running water. Cook food immediately after defrosting.

- Use caution when serving food. Throw out any leftovers that have been at room temperature for more than two hours or in hot weather for more than an hour. If cold food needs to sit out for longer than two hours, use a tray of ice under the food to keep it cold. If hot food must sit out for longer than two hours, use warming trays to keep the food hot.
- Throw it out when in doubt. If you are not sure if a food has been prepared, served or stored safely, throw it away.
- Know when to avoid certain foods altogether.

Interactions

Food contamination usually causes abdominal discomfort and pain, and diarrhea, but symptoms vary depending on the type of infection. Transmission usually occurs via the fecal/oral route with the ingestion of the pathogen present in the contaminated food. After they are ingested, there is a delay, (incubation period) before symptoms appear, that may range from hours to days, depending on the organism. During this period, the microbes pass through the stomach into the intestine, where they start to multiply. Some types stay in the intestine, others produce a toxin that is absorbed into the bloodstream, and others can directly invade the deeper body tissues. The symptoms depend on the type of infection. Numerous pathogens cause similar symptoms, for instance diarrhea, abdominal cramps, and nausea.

Aftercare

There are many different kinds of foodborne diseases and they may require different treatments, depending on the symptoms they cause. Illnesses that cause diarrhea or vomiting lead to dehydration if the person loses more body fluids and salts (electrolytes) than they take in. Replacing the lost fluids and electrolytes is therefore important. If diarrhea is severe, oral medication such as Ceralyte, Pedialyte or Oralyte, can be taken to replace the fluid losses. Preparations of bismuth subsalicylate, such as Pepto-Bismol, can help reduce the duration and severity of the diarrhea.

Complications

Food poisoning is especially serious and potentially life-threatening for young children, pregnant women and their fetuses, older adults, and people with weakened immune systems.

Data from the Centers for Disease Control and Prevention's (CDC) FoodNet surveillance system suggests that although younger individuals usually face far higher rates of infection from foodborne pathogens, older adults, along with the very young are more likely to have severe complications from these infections. In particular, research has shown that the elderly are more vulnerable to gastroenteritis-induced deaths. It is also estimated that 2–3% of all acute foodborne illnesses develop secondary long-term illnesses and complications called chronic sequellae. These can occur in any part of the body, such as the joints, nervous system, kidneys, or heart.

Parental Concerns

A bottle-fed infant is at higher risk for severe infections with bacteria that can grow in a bottle of warm formula if it is left at room temperature for many hours. Particular care is needed to keep baby bottles clean and disinfected. Leftover milk formula or juice should also not be kept in the bottle for many hours

Biological Contamination

Biological contamination generally refers to contamination of our food or environment with microorganisms. This means bacteria, viruses, fungi, and parasites.

Conditions Promoting Biological Contamination

Ambient Conditions

Very low temperatures tend to inhibit the growth of many organisms that readily replicate at room temperature. However, the yeast Sporobolo - myces and the mold Aureobasidiu R*!!•!* (!- lularia) grow and sporulate well in cool environments. Some microbes thrive at higher temperatures. While the fungal genus Aspergillus can grow between 12° and 57 °C, the optimal temperature range for growth is between 37° and 43 °C (body temperature). The bacterium Legionella pneumo- phila and the Actinomycetes Faeni rectivirgula (formerly Micropolyspora faeni) and Thermoacti- nomyces vulgaris grow best at temperatures in excess of 50 °C.L. pneumophila can grow in water temperatures of up to 140 °F and even reside and survive within protozoa. Higher temperatures also increase the respiratory rate of organisms and thereby increase the load of contaminants from metabolism such as water vapor and CO_2.

When the temperature indoors exceeds the outdoor temperature, hot air inside the building rises in a column, creating a positive pressure above. This positive pressure forces the air through available openings (leaks) in the top of the structure, reducing air pressure at the bottom of the building, and thereby increasing the infiltration of outdoor air to equalize the outdoor/indoor pressure gradient. If the air immediately outside the building is contaminated by biological contaminants, the net result of this “stack effect” is to increase indoor air contamination.

High relative humidity promotes the growth of many molds, bacteria, and the arachnid dust mite by providing an abundance of one required nutrient, water. In general, maintaining an indoor relative humidity between 35% and 50% will minimize condensation and indoor dampness, reduce the growth of fungi, dust mites, and bacteria, and provide a reasonable comfort level for the building's occupants.

Outdoor wind currents can transport contaminants from long distances and also stir up and aerosolize many biological contaminants that inhabit ground structures and soil. Kozak found that the Santa Ana winds (from the eastern desert) in Southern California increased the outdoor viable mold spore count in the coastal areas from a usual baseline of 1000 to 1500 spores/m^3 to 43,946 spores/m^3 Yardwork, such as disturbing compost piles or mowing the lawn, can increase outdoor mold counts more than 1000-fold and potentially increase the pollen concentration in the outdoor air of those plants that are disturbed while pollinating. The wind can also increase the pressure gradient between the outdoor and indoor environment, driving indoors the contaminants in the outdoor air.

Because the outdoor air is a rich source of biological contaminants, the availability of ports of entry into the building becomes important, especially if a building is tight (little air leaks in or out). Most commercial buildings constructed in the past 20 years have been built "tight" to conserve energy and depend on mechanical ventilation to supply fresh air and exhaust stale indoor air. If the HVAC (heating, ventilating, and air-conditioning) system does not provide adequate fresh air, air filtration, and exhausting of contaminants, it may actually concentrate outdoor and indoor contaminants within the building. Also, larger and denser bioaerosols tend to settle out of the air more readily than smaller, less dense particles, which tend to remain suspended longer. As a result, indoor environments with better ventilation generally are more effective clearing the air of small respirable bioaerosols.

The degree and type of light can affect biological contaminant growth. Ultraviolet light inhibits the growth of many bacteria and some molds. found that marked shading proximate to a house increased the indoor mold spore counts fivefold. Yet a total absence of light will inhibit the sporulation of some molds, such as Alternaria and Drechslera.

Nutrient Sources

Generally, the most limiting and significant nutrient source is the presence of dead organic matter, even though some molds, many bacteria, and all viruses grow and replicate on or within living substrates. Molds and bacteria typically seed and grow on organic debris found in soil, compost, and dung heaps, wood piles, hay, animal feed, and dead plants or leaves. They can also grow on building and finishing materials, especially if there has been wetting or water damage. Paint, wallpaper, carpeting, especially those with jute (a plant fiber) backing, foam rubber carpet pads, drapery, upholstery fabric and filler, soap scum on bathroom tile and the bathtub, baseboards, hardwood floors, ceiling tiles, cement, exterior and interior wood beams, and framing and roofing material can provide nutrients sufficient to grow bacteria and fungi. Exposed sound and thermal insulation within walls and ventilation ducts provides an organic substrate for microorganism growth. Also, dust and dirt accumulation within insulation material or within ventilation ducts and other HVAC components add further substrate. Disrupting or disturbing contaminated structural components, such as occurs with renovation or repair, can increase the concentration of indoor air contaminants 1000-fold.

Table: Water or moisture reservoirs in or around buildings.

Outdoors	Swamp coolers	Drainage ditches
Cooling towers, small lakes fountains. Poorly kept landscaping Industrial HVAC exhaust. Wells or potable water storage containers Sewer drains. Indoors. Humidifiers Leaky appliances. Subterranean rooms. Leaky roofs. Water-damaged carpet, ceiling tiles, walls, and furniture.Bathroom showers, tubs, carpet, wallpaper, and window coverings. Condensation on windows, cold water pipes, ventilation ducts, and insulation.	Agricultural storage (e.g., grains) Air conditioners. Wood piles. Under house crawlspaces. Air conditioners Leaky plumbing. Ground-level cement slabs and walls Improperly placed vapor barriers.	Compost piles Wooden structures. Water softeners. Crawlspaces . Refrigerator pans.

Stagnant water from wells and hot water heaters, humidifiers, vaporizers, and condensate pans of HVAC systems and cooling units (e.g., water towers, air-conditioners, and evaporative coolers) can harbor bacteria, fungi, algae, and protozoa. Showers, water particle emitters (e.g., vaporizers and humidifers), and ventilation systems can aerosolize water particles adsorbed with live organisms or pathogenic portions of the organisms. Table provides a comprehensive list of potenial water or moisture reservoirs that can be found in or around buildings. These microbes may be living within the water source or on surfaces over which the water or air currents highly saturated with water vapor droplets move. Alternatively, the water or air currents may dislodge nearby biological contaminants and sweep them up, disseminating them in the air or water stream (including potable water) a long distance from their nutrient source. Even in dry environments, focal sources of moisture or water can result in significant biocontaminant growth. Locating nutrient sources and recognizing the ambient climatic conditions can suggest the type of biocontaminants that may be present.

Table: Potential sources of indoor air biological contaminants.

Carpets	Bacteria, fungi, and protozoa
Plants, animals, birds,	Standing water
and humans	Humidifiers, and
Pillows, bedding,	evaporative coolers
and house dust	Hot water tank
Wet or damp material	

Microbial Source

Table lists some potential sources of biological contaminants. Pet excreta and urine provide nutrient sources for bacteria and fungi. They also contain proteins that can cause immunologic diseases; for example, pigeon and parakeet droppings can produce hypersensitivity pneumonitis (pigeon breeder's disease). Cat excreta may contain Toxoplasma gandii and bird excreta Cryptococcus neoformans and Histoplasma capsulatum. Pets' saliva, hair, dander, feathers, or urine are composed of animal proteins that can be allergenic and may persist for years after removal of the pet. Fecal material from dust mites and cockroaches contain very potent allergens, and Legionella may grow within the cysts of amoebae that protect them from the germicidal effects of chlorination of the water source. Kozak et al.5 found that the most prevalent fungal spores within homes in sites evaluated in Southern California, in descending order, were Cladosporium (100% of homes), Penicillium (91%), nonsporulating mycelia (90%:), and Alternaria (87%).

Induction of Human Illness by Biologic Contaminants

The ability of airborne particles to reach different parts of our respiratory tract depends on their size. Large particles in the size range of 30 to 60 Iim usually consist of organic and nonorganic dirt, fibers, and the larger pollens and mold spores. These are filtered out by the nasal vibrissae. Many pollens, mold spores, hyphal fragments, and smaller inert particles (some containing adsorbed biological contaminants) in the range of 5 to 20 jIm will impact on the nasal mucosa or penetrate further down into the major lower airways, that is, primary and secondary bronchi. Bacteria, smaller fungal spores, and droplet nuclei from talking, coughing, and sneezing make up the group of truly respirable particles, that is, those reaching the terminal airways and ranging in size from 1 to 5 m. Particles smaller than 1 im are generally expelled from the human respiratory tract with exhalation. Although viral particles fall into this category, they usually enter the respiratory tract adsorbed to droplet nuclei that may remain there, eventually penetrating the respiratory tract epithelial cell in cases of successful infection.

Biological contaminants most often disseminate in the indoor environment through air currents or water aerosols. The great majority of biological contaminants cause human illness or discomfort by three mechanisms: (1) infection, (2) intoxication, and (3) immunologic responses. Infectious bacteria, viruses, and some fungi enter the human host through the mucous membranes of the respiratory tract or eye, through disrupted areas of our skin (e.g., atypical mycobacteria), and some times through ingestion (e.g., Salmonella). Occasionally infection is iatrogenic from use of contaminated parenteral products such as blood. The microbial load (concentration) is an important determinant of the probability and eventual severity of infection. Some of these infectious agents are saprophytes, which are usually only hazardous for the immunocompromised or debilitated host. Aspergillus, Candida, Geotrichium, Scedosporium, Paecilomyces, and Scopulariopsis are examples of saprophytic mold spores that are small enough to enter the lower airways and grow well at 37 °C. However, under certain conditions, they can become pathogenic, resulting in infection or hypersensitivity reactions. The desquamated skin of occupants of a building is the predominant source of bacteria normally found indoors e.g., Micrococcus and Staphylococcus. Our respiratory tract provides an additional source of ambient bacteria and sometimes viruses.

Nosocomial organisms can spread indoors and cause human diseases, such as the anthrax bacillus contracted from the indoor processing of animal products from infected animals and Q-fever, a rickettsial organism usually contracted from infected laboratory animals or in buildings nearby areas where animals are housed. Two episodes in which tuberculosis bacillus dissemination has been linked to ventilation systems' are of great concern given the recent epidemic of tuberculosis in this country. Viruses may be spread by unconventional routes and cause infection: A measles epidemic within an elementary school spread by its HVAC system and rabies contracted by inhalation of rabies virus in a bat cave harboring infected bats. An unusual situation arose in England where Acanthomoeba infections of the eye occurring in persons wearing contact lenses were linked to the use of contaminated tap water to make saline from salt tablets.

Biological contaminants rarely cause intoxication. A few molds synthesize mycotoxins as secondary metabolites. Mycotoxins are highly variable complex polypeptides that are generally not volatile and remain associated with fungal structures, including spores, or dissolved within the substrate on which the model is growing. Many genera of fungi can produce mycotoxins, but this toxigenic potential is species specific. For example, Aspergillus flavus, Penicillium viridicatum, and Stachybotrys atra are frequent mycotoxin producers, whereas Aspergillus fumigatus and Penicillium chrysogenum are not. The environmental conditions also help determine whether a toxic genic fungus will actually produce mycotoxin. Mycotoxins, when ingested, can produce central nervous system effects (anorexia, nausea, and fatigue), immunosuppression, gastrointestinal lesions, hematopoietic suppression, and suppression of reproductive function. Some, such as the aflatoxin produced by A. flavus, are potent carcinogens. Because mycotoxins are present in high concentrations in some mold spores, many health experts share the opinion of the U.S. Environmental Protection Agency that "it is reasonable to assume that these toxins have a systemic effect when inhaled, since inhalation more effectively allows entry for dissolved substances. However, to date despite several reports linking inhaled mycotoxins to human illness, few well-documented cases of inhalation-induced human mycotoxicosis exist. One such case involved a house heavily infested with S. atra.

In addition to toxins produced by fungi, may gram-negative bacteria produce endotoxin, a potent lipopolysaccharide moiety of the cell wall. This toxin produces fever, malaise, respiratory distress, peripheral blood leukocytosis with a shift to the left, and shock, which can be fatal. However, well-defined reports of human intoxication from inahled endotoxin have been limited to experimentally induced air flow obstruction. Immunogenic substances that are integral parts of the structures of various plants, animals, and microorganisms or that are released by them into the environment can cause indoor environmentally induced human illness. These immunogenic epitopes can produce immunologically induced inflammation by all four Gell and Coombs pathologic mechanisms: (1) immediate (IgE-mediated) hypersentivity or allergy, (2) cytotoxic antibodies against homologous tissue antigens (e.g., group B hemolytic streptococcus and rheumatic fever), (3) antibody-antigen complex disease (e.g., precipitins in hypersensitivity pneumonitis), and (4) delayed-type T cell-mediated immunity (e.g., Mycobacterium tuberculosis).

A lack of understanding of several different aspects of these immunologic responses can lead to confusion or misunderstanding regarding the relevance of environmental biocontaminants to human illness. Our ability to distinguish different mechanisms of inflammation is blurred somewhat because often more than one mechanism may contribute to the pathologic findings. For example, immediate hypersensitivity involves both cell-mediated and humoral immune activity. Also, antibody responses, including IgE, can be mea sured against antigens to infectious agents such as viruses that have traditionally been thought of as evoking cell-mediated immunologic responses. The exact role of antiviral (e.g., respiratory syncytial virus) IgE in the pathogenesis of asthma remains unclear. Microbial antigens from fungi, bacteria, and protozoa can induce specific precipitating antibodies (IgG, IgM, and IgA) in susceptible persons. A role for local immune complexes in the pulmonary parenchyma or airways has been suggested in the pathogenesis of hypersensitivity pneumonitis' - and allergic bronchopulmonary aspergillosis.' Cytotoxic antibodies may also contribute to the pulmonary parenchyma damage seen in hypersensitivity pneumonitis.

Hypersensitivity responses to biological components are limited to those with the genetic ability to respond to specific allergens, such as an IgE response to the reproductive progeny of plants (pollen) and fungi (spores), the excreta of insects and arachnids (cockroach and dust mite), structural components and byproducts of animals and insects (danders, saliva, urine, feathers, and venom), and products produced from some plants (pyrethrins used in insecticides, cotton linters, kapok, and gums).

Bacteria, such as Bacillus subtilis, can synthesize enzymes that stimulate specific IgE synthesis in a small minority of genetically predisposed people who become sensitized to these allergens. There have been several reports of occupational asthma in detergent manufacturing workers and in other occupations involving frequent large exposures to this and other enzymes. Prior exposure and the nature of this exposure, that is, high or low dose, continuous or intermittent, may play a determining role in whether illness develops and what form it will take in a given individual.

Indoor air may not be the vehicle by which certain pathogenic biocontaminants contact those who become ill. Plant resins from poison oak, ivy, and sumac can be spread in the indoor environment from contaminated clothing or human contact to the skin of family members. Individuals sensitive to these allergens can develop contact dermatitis to these resins and to other natural plant products (some of which may later become airborne, e.g., ragweed pollen) that can mimic photosensitivity dermatitis or atopic eczema. Once the inflammation becomes established, the inital immunogens may no longer be required to perpetuate the inflammatory response, which then becomes non-specific and may persist for months to years.

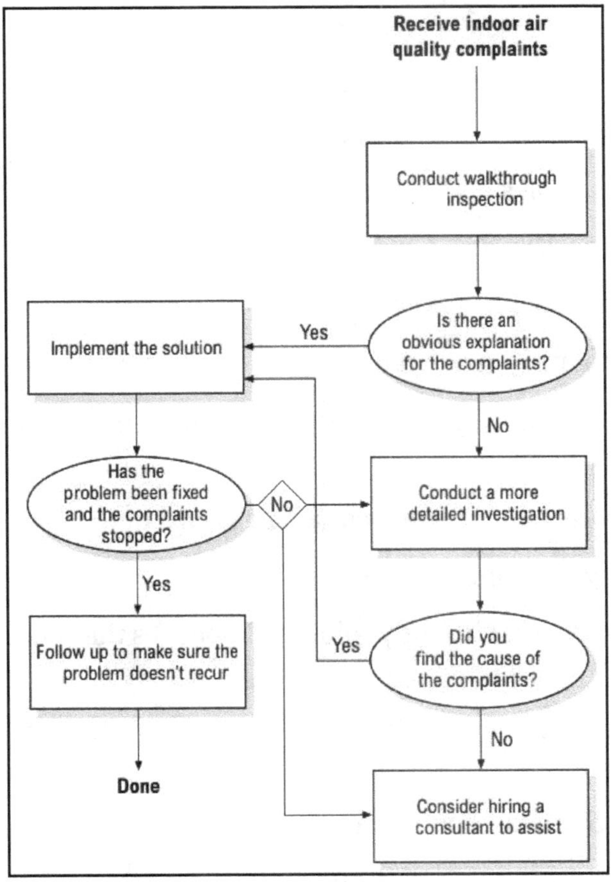

Flowchart of an investigation strategy for indoor air quality problems.

Some molds and indoor house plants synthesize and release volatile organic compounds into the environment, which can result in a characteristic and sometimes irritating odor, such as the musty odor from mold overgrowth. Whether volatile organic compounds of organic origin pose significant human health hazards other than the discomfort caused by their odor remains as yet undetermined.

Physical Contamination

Physical contamination occurs when a physical object enters food at some stage of the production or preparation process. Physical objects in food can be a choking hazard and often introduce biological contaminants as well. Even if the object is not likely to injure your customer, finding an object in their food can be very distressing for a customer (who knows that harmful microorganisms on the object could make them ill).

Common examples of physical contaminants in food businesses include:

- Hair,
- Fingernails,
- Bandages,
- Jewellery,
- Broken glass, staples,
- Plastic wrap/packaging,
- Dirt from unwashed fruit and vegetables,
- Pests/pest droppings/rodent hair.

To minimize the risk of physical food contamination occurring in your food business, always:

- Wear hair neatly tied back or wear a hair/beard net,
- Keep jewellery to a minimum,
- When necessary, wear brightly coloured bandages that can be easily seen if they fall off,
- Throw out and replace cracked, chipped or broken dishware, glassware and equipment,
- Use a plastic or metal scoop for ice (never use the glass!),
- Wash fruits and vegetables thoroughly,
- Establish pest prevention and control procedures as part of your Food Safety Plan.

Examples:

Glass Fragments

X-ray microanalysis in the scanning electron microscope is used to gain an elemental composition of a glass fragment. This composition is then interpreted to identify a specific type of glass such as soda-lime glass, heat-resistant glass and float glass.

Plastic

FT-IR spectroscopy is performed on plastic samples to gain a spectrum of infrared absorption. This is then compared against a vast database of reference spectra to identify the material. Some commonly identified plastics are PVC, polyethylene, polycarbonate, polysulfone, polypropylene and polystyrene.

Metal

X-ray microanalysis in the scanning electron microscope is used to gain an elemental composition of metal. The suite of elements within a sample can distinguish between different types of metal such as steel, stainless steel, stainless steel type 316, nickel-plated steel, galvanised steel and corroded steel.

Bone

Suspected bone samples can be confirmed by examination under the compound microscope for pore-like structures consistent with bone, known as lacunae. This is followed by X-ray microanalysis in the scanning electron microscope to determine the elemental composition of the sample. Large peaks of phosphorous and calcium within the spectrum are indicative of bone.

Pharmaceuticals

Suspected pharmaceutical samples can be analysed by our chromatography team using gas chromatography and liquid chromatography to look for peaks of interest which may be consistent with an active ingredient, such as amoxicillin which is used in antibiotics.

The microscopy section can also use FT-IR spectroscopy in combination with an identification database and compound microscopy to try and determine the type of capsule shell and in some cases the presence of chemical compounds, fillers and active ingredients found within pharmaceutical products.

Insects

Suspected insects can be examined under the stereo microscope to identify features such as legs, antennae and wings. Samples which require speciation can then be sent off for further analysis to expert entomologists at one of our approved subcontractors, who will aim to provide an identification. In some cases, the experts are able to advise on typical geographical distributions and life cycles of insects. A chemical test known as a phosphatase test can also be carried out by the microscopy section, which aims to identify whether the phosphatase enzyme within a once-living sample, such as an insect, is still active or if it has been denatured though heat treatment/cooking. Please note, samples which are mouldy may not be suitable for phosphatase analysis as the mould will also naturally produce the phosphatase enzyme and may give a false positive result.

Plant Matter/Wood

Sectioning, staining and examination under the compound microscopeof suspected plant matter can reveal cell types and structures which may be indicate a specific source. Wood samples can also be sectioned, stained and examined under the compound microscope, which can reveal features consistent with various species of hardwoods, softwoods and woody monocots. In addition, we have a reference collection of woods within our database which can assist with comparing features seen within complaint samples with those found in known samples to identify a likely source.

Animal Tissue

Sectioning, staining and examination under a compound microscope is used to identify features consistent with tissues from an animal such as muscle fibres, adipose tissue, elastin fibre, collagen and cartilage. In some cases, FT-IR spectroscopy is also used to gain an infrared absorption spectrum of a sample, which may produce a spectrum consistent with organic matter with bands consistent with protein.

Silicate Minerals

X-ray microanalysis is used to determine the elemental composition of the sample. Silicate minerals will contain high levels of silicon and oxygen. In some cases, a large peak of aluminium will also be seen, indicating an alumino-silicate mineral.

Hair

Features common to hairs are assessed under a compound microscope(and used to identify potential sources). These features include the scale pattern, the medulla and the shape of the root, tip and shaft. These structures can aid in the identification of the hair as human (and the potential location on the body) or animal and if animal the possible species.

Chemical Contamination

The phrase chemical contamination is a clear indication of the presence of chemicals where they should not be or are present in an amount that is in a higher concentration than the amount that is attributed as safe. The chemical hazards are one of the main causes of food contamination that associated with foodborne disease outbreaks.

Chemical contaminants may occur in our food from various sources. They typically pose a health concern, resulting in strict regulations of their levels by national governments and internationally by the Codex Alimentarius Commission. Therefore, analysis of relevant chemical contaminants is an essential part of food safety testing programs to ensure consumer safety and compliance with regulatory limits. Modern analytical techniques can determine known chemical contaminants in complex food matrices at very low concentration levels. Moreover, they can also help discover and identify new or unexpected chemical contaminants.

Sources of Chemical Contaminants in Food

Chemical contaminants can be present in foods mainly as a result of the use of agrochemicals, such as residues of pesticides and veterinary drugs, contamination from environmental sources (water, air or soil pollution), cross-contamination or formation during food processing, migration from food packaging materials, presence or contamination by natural toxins or use of unapproved food additives and adulterants.

Pesticide Residues

The use of pesticides, such as insecticides, fungicides or herbicides, has become an integral part of modern agriculture to increase crop yields and quality by controlling various pests, diseases and weeds. Registration of new pesticides is a strictly regulated process that evaluates their toxicity and environmental fate, and sets maximum residue limits (tolerances) in raw and processed commodities. There are over 1,400 known pesticides. Some of them should no longer be used but may still be present in the environment. Older pesticides are being reevaluated based on currently available scientific data.

Approved uses of pesticides following Good Agricultural Practices should result in pesticide residues below maximum residue limits established in a given country. However, global sourcing of raw commodities and global distribution of food products complicate the situation because pesticide registrations, uses and limits can be and are different in different countries. Consequently, an approved use in one country may result in an illegal pesticide residue in a food imported into another country, such as the recent case of the fungicide carbendazim in orange juice imported into the United States from Brazil. Furthermore, pesticides can be misused or present in food due to contamination during application (spray drift), storage or transportation or from environmental sources, such as contaminated water or soil.

Veterinary Drug Residues

Similar to pesticides, veterinary drugs are agrochemicals that undergo a thorough registration process, resulting in setting of their maximum residue limits/tolerances in animal-derived foods. The major classes of veterinary drugs include antibiotics, anthelmintics, coccidiostats, nonsteroidal anti-inflammatory drugs, sedatives, corticosteroids, beta-agonists and anabolic hormones. These drugs, which are administered to live animals, can remain as residues in animal tissues. Liver and kidney are highly susceptible to residues given their biological function.

Certain antibiotics, such as penicillin, can cause severe allergic reactions in sensitive individuals, which is an important reason for enforcing their residue limits in foods of animal origin. Another important justification for limiting antibiotic usage in food-producing animals is to reduce the risk of pathogenic microorganisms becoming resistant to antibiotics. Most veterinary drugs are not of acute toxicological concern, but some substances, such as nitrofurans, chloramphenicol, clenbuterol and diethylstilbestrol, have been banned in most countries due to their carcinogenicity. Concern about endocrine-disrupting effects has become another reason for regulation of certain veterinary drugs, such as beta-agonists and hormones.

Environmental Contaminants

Environmental contaminants can be man-made or naturally occurring substances present in air, water or soil. They can enter the food chain and even bioaccumulate. Some can pose an acute health risk if present at higher concentrations, but the major concern related to the presence of environmental contaminants in foods is their potential endocrine disruption, developmental, carcinogenic and other chronic effects.

Examples of environmental contaminants that enter the food chain include heavy metals, polychlorinated biphenyls (PCBs), "dioxins" (polychlorinated dibenzodioxins and dibenzofurans), persistent chlorinated pesticides (e.g., DDT, aldrin, dieldrin, heptachlor, mirex, chlordane), brominated flame retardants (mainly polybrominated diphenyl ethers), polyfluorinated compounds, polycyclic aromatic hydrocarbons (PAHs), perchlorate, pharmaceutical and personal care products or haloacetic acids and other water disinfection byproducts.

The manufacture and use of PCBs and other persistent organic pollutants (POPs) have been banned for years, but they remain in the environment due to their high stability. PAHs can be found in the environment as a result of industrial pollution or can originate from oil spills; thus, they were of concern in seafood after the oil spill accident in the Gulf of Mexico in 2010.

Food Processing Contaminants

Certain toxic or undesirable compounds can be formed in foods during their processing, such as during heating, baking, roasting, grilling, canning, hydrolysis or fermentation. Precursors of these contaminants can occur naturally in the food matrix, such as in the case of acrylamide being formed during the Maillard reaction between the amino acid asparagine and a reducing sugar (especially in potato- and cereal-based, heat-treated products). Alternatively, certain processing contaminants, such as nitrosamines, can be formed by interaction of natural food components with food additives. Carcinogenic and genotoxic chlorpropanols, such as 3-monochloropropane-1,2 diol (3-MCPD), are formed during the acid hydrolysis of wheat, soya and other vegetable protein products.

Examples of other processing contaminants include PAHs (in grilled and smoked products), ethyl carbamate (in yeast-fermented alcoholic beverages and other products) or furan (in a variety of heat-treated foods, especially coffee and canned/jarred food).

Food processing may also be a source of cross-contamination, such as contamination of nonallergenic foods with known food allergens.

Migrants from Packaging Materials

Direct contact of foods with packaging materials can result in chemical contamination caused by migration of certain substances into foods. Examples of migrants of health concern may include bisphenol A or phthalates from plastic materials, 4-methylbenzophenone and 2-isopropylthioxanthone from inks, mineral oil from recycled fibers or semicarbazide from a foaming agent in the plastic gaskets that are used to seal metal lids to glass packaging. Toxins are naturally occurring substances that are produced by various organisms, with mycotoxins and marine biotoxins typically representing the major concerns in foods. Other examples of toxins in foods may include bacterial toxins (e.g., staphylococcal toxins) or certain plant toxins, such as pyrrolizidine alkaloids that can be found in honey, milk or eggs. While the bacterial/fungal contamination can be eliminated with heat treatment, the toxins can remain in the food product as contaminants.

Mycotoxins are toxic secondary metabolites produced by fungi (molds) that can colonize various crops. They are of concern mainly in cereals, nuts, infant formula, milk, dried fruit, baby food, coffee, fruit juice and wine. There are many mycotoxins, but only a few are currently regulated, with the European Union having a more comprehensive list than most other countries, which includes aflatoxins, ochratoxin A, patulin, deoxynivalenol, zearalenone, fumonisins and T-2/HT-2 toxins. Different mycotoxins are prevalent in different climates and in various growing and storage conditions.

Marine biotoxins, such as saxitoxin, domoic acid, okadaic acid or ciguatoxin, are highly toxic compounds produced by phytoplankton. During so-called harmful algal bloom events, they can accumulate in fish or shellfish, such as clams, mussels, scallops or oysters, to levels that can pose serious health risks or even be lethal to humans.

Unapproved Food Additives and Adulterants

Food adulteration can happen accidentally when unapproved additives are introduced to the food, or the wrong additive is introduced through formulation error. This results in mislabeled food. Perhaps a larger health issue is when foods are adulterated intentionally for economic reasons to sell a low-value food or material for more or to mask food spoilage. Some adulteration may just mislead or cheat consumers, such as adding high fructose corn syrup to honey, but some may be harmful to them. The most notorious example from recent years is the addition of melamine to whey and other protein concentrates to increase their apparent protein content analyzed as total nitrogen. Other examples include the use of toxic Sudan dyes in adulterated chili powders or adulteration of virgin olive oil with hazelnut oil, which can cause unexpected allergic reactions in sensitive individuals. Analysis of Known Chemical Contaminants in Food.

Most known chemical contaminants in foods are small organic molecules. Except for high-level adulterants, they are typically present in foods at low concentrations (parts per trillion to parts per million); thus, their analyses in complex food matrices are often quite challenging. The basic analytical approach involves an extraction using a suitable solvent, cleanup to remove interfering matrix components, a chromatographic separation and a selective detection.

It is not an exaggeration to say that the implementation of mass spectrometry (MS) as a detection technique has truly revolutionized the analysis of chemical contaminants in foods. As opposed to element-selective or nonselective detectors, MS can detect a wide range of compounds independent of their elemental composition and provide simultaneous quantitation and structural identification of detected analytes. It also adds another degree of separation/selectivity on top of chromatographic separations. These unique features have made MS the number one choice for detection and identification/confirmation of trace-level organic chemical contaminants in modern testing laboratories.

First, the combination of MS with gas chromatography (GC-MS) has become popular for the analysis of volatile and semivolatile compounds, including many pesticide residues, PAHs, PCBs and other less-polar POPs. More polar, thermolabile and less volatile analytes were difficult to analyze until the more recent introduction of atmospheric ionization techniques, such as electrospray, for liquid chromatography-mass spectrometry (LC-MS). LC-MS has opened the door to the direct analysis of many more polar contaminants, including modern, new-generation pesticides, and the majority of veterinary drugs and toxins, such as mycotoxins. Many of the emerging and recently identified contaminants, including acrylamide, melamine or Sudan dyes, are analyzed preferably by LC-MS.

Thus, modern food contaminant testing laboratories utilize both GC-MS and LC-MS to cover the wide polarity range of possible organic chemical contaminants. Tandem MS (MS/MS) is typically employed to provide an increased selectivity (especially in LC-MS) that helps further distinguish target compounds from potential matrix interference. Figure shows an example of chromatograms obtained in a multiresidue analysis of more than 300 pesticides analyzed by GC-MS/MS and LC-MS/MS, demonstrating the speed and selectivity of state-of-the-art instruments that enable the simultaneous and highly sensitive analysis of many compounds.

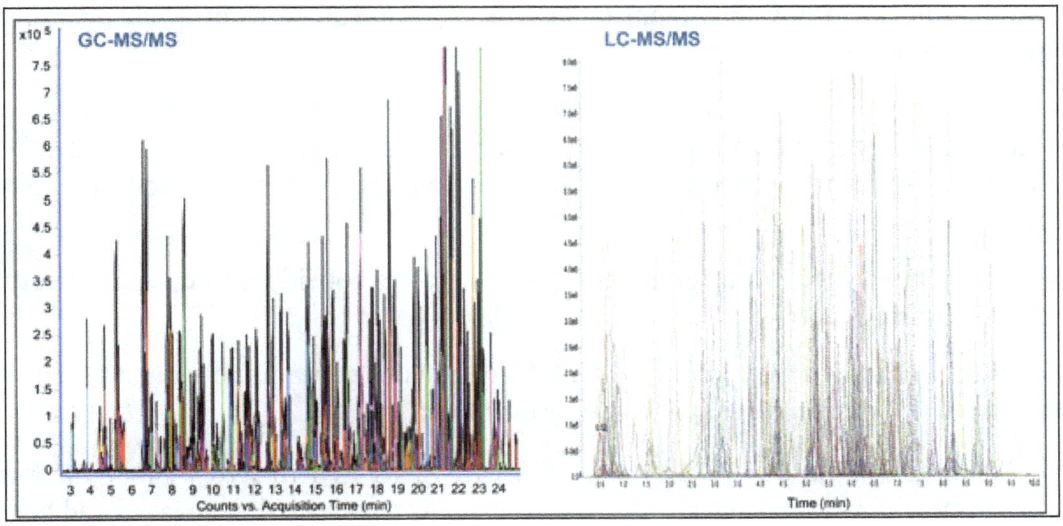

Identification of Unknown Chemical Contaminants in Food

Detection and identification of unknown contaminants is not an easy task, especially if they are present at low concentration levels. It requires expertise and a good analytical strategy that is based on all gathered information about the sample and potential sources of contamination. Any clues, such as changes in smell, taste or texture, as well as a description of potential poisoning symptoms may be important in this respect. Concurrent analysis of control ("good") samples with suspect samples is often essential to find differences and eliminate potential false positives.

If a certain compound or a group of compounds is suspected, then a targeted sample preparation and instrumental method(s) can be employed. For a truly unknown analysis, different extraction and separation approaches should be used to isolate compounds with a wide range of physicochemical properties (polarity, solubility, volatility, etc.). Nontargeted analysis should be performed, such as MS with full-spectra acquisition. Statistical analysis of the acquired chromatographic and MS data of contaminated and noncontaminated samples may help identify differences and reduce the number of components that have to be examined. The acquired MS spectra of suspected contaminants can be compared with MS spectral libraries and compound databases. In LC-MS, high-resolution/accurate-mass measurements, using time-of-flight (TOF) or orbitrap MS instruments, should be used for added selectivity. In addition, tandem MS should be employed to help elucidate the structure of unknown contaminants. In the end, strong knowledge and expertise in both analytical and food chemistry are typically required to succeed in this task.

Cross-contamination

Cross-contamination is the unintentional transfer of microorganisms, chemical contaminants (including allergens) or any foreign substance from food, person, or object to another food product.

It usually occurs from raw foods to ready-to-eat products (RTE) or between products that contain allergens and allergen-free products. Cross-contamination can cause food poisoning when harmful bacteria are transferred to RTE products that do not undergo further processing to eliminate bacteria.

Food, equipment, food contact surfaces and people are considered important sources of cross-contamination.

Food: Raw food which may contain harmful bacteria can contaminate other food products when they come in direct contact. For example:

- Contamination of RTE products with pathogens transferred by blood dripping from raw meat during storage (if RTE products are stored uncovered and below raw meat products).
- Contamination of food products with pathogens transferred from leaky boxes or broken cartons containing raw products during receiving.
- Contamination of RTE products with pathogens during packaging due to the use of contaminated packaging material.
- Contamination of an allergen-free product with residual allergens on the equipment surface due to an ineffective cleaning.
- Contamination of RTE products with residues of chemicals (used to wash and sanitize tables and equipment) and allergens due to improper sanitation procedures.
- Equipment and food contact surfaces: Food residues on equipment and other food contact surfaces may provide opportunities for cross contamination of food products.
- Contamination of RTE products with bacteria and allergens due to improper washing and sanitation of equipment and utensils.
- Contamination of food products with bacteria and allergens due to the use of dirty cloths to clean the surface of tables.
- People: Food workers that do not follow Good Manufacturing Practices (GMPs) may transfer microorganisms and allergens to the food.
- Contamination of food products with pathogens and allergens caused by not following prescribed food handling procedures (ex: taking shortcuts).
- Contamination of RTE products due to soiled uniforms and gloves or dirty boots.

Preventing Cross-contamination

Cross-contamination is usually preventable. To decrease the chances of cross-contamination, keep raw and cooked products separate and allergen foods and allergen-free products separate.

GMP Implementation Recontamination of food products and food contact surfaces are prevented by:

- Washing your hands when starting work, changing tasks, after using the washroom, and after breaks.
- Using hand washing stations with soap and sanitizer (for example, 30-50 ppm chlorine solution).
- Using foot dip stations at the employee entry door.
- Using different utensils for raw and pre-prepared foods.
- Not placing containers or boxes that may have been stored on the floor onto food contact surfaces.
- Cleaning and sanitizing utensils between uses.
- Using correct sanitation procedures for all utensils and equipment.
- Changing smocks and dipping or changing boots before entering a RTE area.

Storage

- Store raw food and RTE food separate.
- Store ingredients containing health priority allergens below or away from allergen-free ingredients (for example, do not store loose bags of nuts over open bin of flour).
- If raw and RTE products are stored in the same refrigerator, store RTE foods above raw foods.
- Keep food products covered and off the floor during storage.
- Store cleaning materials and other chemicals separately from food products.

Plant Design and Production Flow

- Flow of operations in a food processing facility should minimize the likelihood of crosscontamination (for example, employees working in the raw processing area should not access the RTE area).
- Physical separation of raw food products from cooked products is essential.

- Clean filtered air should flow from finished product or packaging areas to raw product handling areas.
- Condensation must be controlled within the processing area; particularly the RTE and packaging rooms.

Cleaning and Sanitation

- Use color-coded cleaning utensils for different sections of the plant.
- Use brushes to clean food contact surfaces.
- If you use cloths to wipe food contact surfaces, dip the cloth in sanitized water between uses.

Training

- Employees must be trained in GMPs.
- Employees should be able to identify a potential cross contamination situation and prevent it from happening.
- Visitors should be educated to ensure that they are following proper GMPs.

Food Borne Illness

Food poisoning, also called foodborne illness, is illness caused by eating contaminated food.

Myriad microbes and toxic substances can contaminate foods. There are more then 250 known foodborne diseases. The majority are infectious and are caused by bacteria, viruses, and parasites. Other foodborne diseases are essentially poisonings caused by toxins, chemicals contaminating the food. All foodborne microbes and toxins enter the body through the gastrointestinal tract and often causes the first symptoms there. Nausea, vomiting, abdominal cramps and diarrhea are frequent in foodborne diseases.

Many microbes can spread in more than one way, so it may not be immediately evident that a disease is foodborne. The distinction matters, because public health authorities need to know how a particular disease is spreading to take the appropriate steps to stop it. For example, infections with Escherichia coli O157:H7 (E. coli O157:H7) can be acquired through contaminated food, contaminated drinking water, contaminated swimming water, and from toddler to toddler at a day care center. Depending on which means of spread cause a case, the measures to stop other cases from occurring could range from removing contaminated food from stores, chlorinating a swimming pool, or closing a child day care center.

The most common foodborne infections are caused by three bacteria - Campylobacter, Salmonella, and E. coli O157:H7 - and by a group of viruses called calicivirus, better known as Norwalk-like virus:

- Campylobacter - Campylobacter is the most common bacterial cause of diarrheal illness in the world. The bacteria live in the intestines of healthy birds, and most raw poultry meat has Campylobacter on it. Eating undercooked chicken, or other food that has been contaminated with juices dripping from raw chicken is the most frequent source of this infection. Aside from diarrhea, common symptoms include causes fever, diarrhea, and abdominal cramps.
- Salmonella - Salmonella is widespread in the intestines of birds, reptiles and mammals. People can acquire the bacteria via a variety of different foods of animal origin. The illness it causes is called salmonellosis and typically includes fever, diarrhea and abdominal cramps. In persons with poor underlying health or weakened immune systems, Salmonella can invade the bloodstream and cause life-threatening infections.
- E. coli O157:H7 - E. coli O157:H7 has a reservoir in cattle and other similar animals. Illness typically follows consumption of food or water that has been contaminated with microscopic amounts of cow feces. The illness it causes is often a severe and bloody diarrhea and painful abdominal cramps, without much fever. But in 3 to 5% of cases, a life-threatening complication called the hemolytic uremic syndrome(HUS) can occur several weeks after the initial symptoms, resulting in anemia, profuse bleeding, and kidney failure.
- Calcivirus (Norwalk-like virus) - Calicivirus (Norwalk-like virus) is an extremely common cause of foodborne illness (though it is rarely diagnosed, because the laboratory test is not widely available). It causes an acute gastrointestinal illness, usually with more vomiting than diarrhea, that resolves within two days. Unlike many foodborne pathogens that have animal reservoirs, it is believed that Norwalk-like viruses spread primarily from one infected person to another. Infected kitchen workers can contaminate a salad or sandwich as they prepare it, if they have the virus on their hands. Infected fishermen have contaminated oysters as they harvested them.

Common diseases that are usually transmitted by other routes are occasionally foodborne. These include infections caused by Shigella, hepatitis A, and the parasites Giardia lamblia and Cryptosporidia.

Food toxins - Some foodborne diseases are caused by a toxin in the food that was produced by a microbe in the food. For example, staph bacteria (Staphylococcus aureus) can grow in food and produce a toxin that causes intense vomiting. The rare but deadly disease botulism occurs when the bacterium Clostridium botulinum grows and produces a powerful paralytic toxin in foods. These toxins can produce illness even if the microbes that produced them are no longer there.

Other foodborne diseases - Among the many other foodborne diseases are the following: amebiasis (Entamoeba histolytica infection), Blastocystis hominis infection, bovine spongiform encephalopathy (BSE) and Creutzfeldt-Jakob disease (CJD), cholera, cryptosporidiosis (crypto), cyclospora cayetanensis, diarrheagenic Escherichia coli (E. coli), viral gastroenteritis, giardiasis, listeriosis, marine toxins shigellosis, travelers' diarrhea, trichinosis (trichinellosis), typhoid, Vibrio parahaemolyticus and Vibrio vulnificus infection.

Magnitude of the problem - An estimated 76 million cases of foodborne disease occur each year in the US alone. The great majority of these cases are mild and cause symptoms for only a day or two. Some cases are more serious, and CDC estimates that there are 325,000 hospitalizations and 5,000 deaths related to foodborne diseases each year in the US. The most severe cases tend to occur in the very old, the very young, those who have an illness already that reduces their immune system function, and in healthy people exposed to a very high dose of an organism.

Prevention and Control

Major foodborne illnesses, including salmonellosis and *E. coli* infections, are reportable diseases in many places, meaning that infections caused by those agents must be reported by physicians and medical laboratories to local, state, or national health departments. However, because most cases of foodborne illness are mild and are not diagnosed, the reported number of cases is assumed to be an undercount of the true number of cases.

Many laws regulating the production, transport, and preparation of food are intended to prevent foodborne illness and limit its consequences. For example, laws have been implemented to help prevent the contamination of raw food, to mandate its safe preparation and storage, and, if necessary, to close restaurants or food suppliers responsible for disease outbreaks or who fail to follow safe food-hygiene practices. There are many means by which raw food may be contaminated, including irrigating or washing with unclean water, contamination of meat and poultry with fecal matter during the slaughtering and packaging processes, preparation by food handlers who carry bacteria or viruses on their hands, or using utensils and preparation surfaces that are not clean.

Cooking at a sufficient temperature kills many microbes and parasites. However, some microbes may also be present in cooked food, such as when contamination occurs during handling after cooking or pasteurization. In some instances, contamination post-cooking or post-pasteurization may not pose a health risk, if only a small number of microbes are present. Most bacteria grow rapidly at room temperature, whereas refrigeration or freezing keeps them from multiplying (L. monocytogenes is a notable exception). Hence, even cooked foods must be promptly refrigerated in order to prevent the multiplication of disease-causing organisms.

Other Foodborne Diseases

Amebiasis (Entamoeba Histolytica Infection)

Amebiasis is a parasitic infection of the intestines caused by the protozoan Entamoeba histolytica, or E. histolytica. The symptoms of amebiasis include loose stool, abdominal cramping, and stomach pain. However, most people with amebiasis won't experience significant symptoms.

Risk of Amebiasis

Amebiasis is common in tropical countries with underdeveloped sanitation. It's most common in the Indian subcontinent, parts of Central and South America, and parts of Africa. It's relatively rare in the United States.

People with the greatest risk for amebiasis include:

- People who have traveled to tropical locations where there's poor sanitation.
- Immigrants from tropical countries with poor sanitary conditions.
- People who live in institutions with poor sanitary conditions, such as prisons.
- Men who have sex with other men.
- People with compromised immune systems and other health conditions.

Causes of Amebiasis

E. histolytica is a single-celled protozoan that usually enters the human body when a person ingests cysts through food or water. It can also enter the body through direct contact with fecal matter.

The cysts are a relatively inactive form of the parasite that can live for several months in the soil or environment where they were deposited in feces. The microscopic cysts are present in soil, fertilizer, or water that's been contaminated with infected feces. Food handlers may transmit the cysts while preparing or handling food. Transmission is also possible during anal sex, oral-anal sex, and colonic irrigation.

When cysts enter the body, they lodge in the digestive tract. They then release an invasive, active form of the parasite called a trophozite. The parasites reproduce in the digestive tract and migrate to the large intestine. There, they can burrow into the intestinal wall or the colon. This causes bloody diarrhea, colitis, and tissue destruction. The infected person can then spread the disease by releasing new cysts into the environment through infected feces.

Symptoms of Amebiasis

When symptoms occur, they tend to appear 1 to 4 weeks after ingestion of the cysts. According to the Centers for Disease Control and PreventionTrusted Source (CDC), only about 10 to 20 percent of people who have amebiasis become ill from it. Symptoms at this stage tend to be mild and include loose stools and stomach cramping.

Once the trophozites have breached the intestinal walls, they can enter the bloodstream and travel to various internal organs. They can end up in your liver, heart, lungs, brain, or other organs. If trophozites invade an internal organ, they can potentially cause:

- Abscesses,
- Infections,
- Severe Illness,
- Death.

If the parasite invades the lining of your intestine, it can cause amebic dysentery. Amebic dysentery is a more dangerous form of amebiasis with frequent watery and bloody stools and severe stomach cramping.

The liver is a frequent destination for the parasite. Symptoms of amebic liver disease include feverand tenderness in the upper-right part of your abdomen.

Diarrhea

Diarrhea - Loose, watery and possibly more-frequent bowel movements is a common problem.

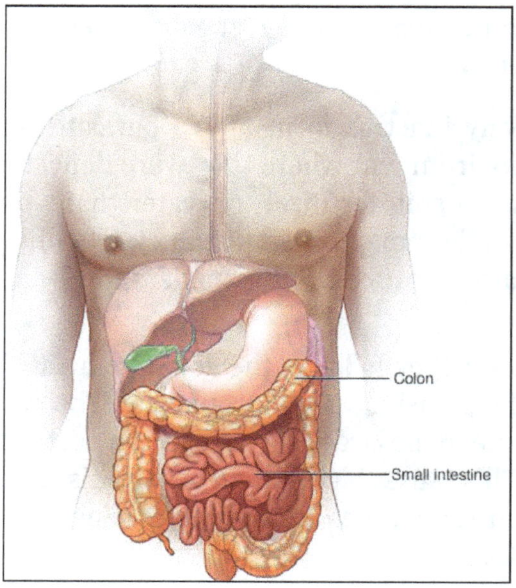

Colon and Small Intestine

The small intestine and colon are components of your digestive tract, which processes the foods you eat. The intestines extract nutrients from the foods. What isn't absorbed by the intestines continues along the digestive tract and is expelled as stool during a bowel movement.

Luckily, diarrhea is usually short-lived, lasting no more than a few days. But, when diarrhea lasts for weeks, it usually indicates that's there's another problem. If you have diarrhea for weeks or longer, you may have a condition such as irritable bowel disorder, or a more serious disorder, such as a persistent infection or inflammatory bowel disease.

Symptoms

Signs and symptoms associated with diarrhea may include:

- Loose, watery stools,
- Abdominal cramps,
- Abdominal pain,
- Fever,
- Blood in the stool,
- Mucus in the stool,
- Bloating,
- Nausea,
- Urgent need to have a bowel movement.

Causes

A number of diseases and conditions can cause diarrhea, including:

- Viruses: Viruses that can cause diarrhea include Norwalk virus, cytomegalovirus and viral hepatitis. Rotavirus is a common cause of acute childhood diarrhea.
- Bacteria and parasites: Contaminated food or water can transmit bacteria and parasites to your body. When traveling in developing countries, diarrhea caused by bacteria and parasites is often called traveler's diarrhea. Clostridium difficile is another type of bacteria that can cause serious infections that cause diarrhea, and it can occur after a course of antibiotics or during a hospitalization.
- Medications: Many medications, such as antibiotics, can cause diarrhea.

Antibiotics destroy both good and bad bacteria, which can disturb the natural balance of bacteria in your intestines. Other drugs that cause diarrhea are cancer drugs and antacids with magnesium.

- Lactose intolerance: Lactose is a sugar found in milk and other dairy products. People who have difficulty digesting lactose have diarrhea after eating dairy products. Lactose intolerance can increase with age because levels of the enzyme that helps digest lactose drop after childhood.
- Fructose: Fructose is a sugar found naturally in fruits and honey. It's sometimes added as a sweetener to certain beverages. In people who have trouble digesting fructose, it can lead to diarrhea.
- Artificial sweeteners: Sorbitol and mannitol artificial sweeteners found in chewing gum and other sugar-free products can cause diarrhea in some otherwise healthy people.
- Surgery: Abdominal or gallbladder removal surgeries can sometimes cause diarrhea.
- Other digestive disorders: Chronic diarrhea has a number of other causes, such as Crohn's disease, ulcerative colitis, celiac disease, microscopic colitis and irritable bowel syndrome.

Complications

Diarrhea can cause dehydration, which can be life-threatening if untreated. Dehydration is particularly dangerous in children, older adults and those with weakened immune systems.

If you have signs of serious dehydration, seek medical help.

Indications of Dehydration in Adults

These include:

- Excessive thirst.
- Dry mouth or skin.
- Little or no urination.
- Weakness, dizziness or lightheadedness.
- Fatigue.
- Dark-colored urine.

Indications of Dehydration in Infants and Young Children

These include:

- Not having a wet diaper in three or more hours.
- Dry mouth and tongue.
- Fever above 102 °F (39 °C).
- Crying without tears.
- Drowsiness, unresponsiveness or irritability.
- Sunken appearance to the abdomen, eyes or cheeks.

Prevention

Preventing Viral Diarrhea

Wash your hands to prevent the spread of viral diarrhea. To ensure adequate hand-washing:

- Wash frequently: Wash your hands before and after preparing food. Wash your hands after handling uncooked meat, using the toilet, changing diapers, sneezing, coughing and blowing your nose.
- Lather with soap for at least 20 seconds: After putting soap on your hands, rub your hands together for at least 20 seconds. This is about as long as it takes to sing "Happy Birthday" twice through.
- Use hand sanitizer when washing isn't possible: Use an alcohol-based hand sanitizer when you can't get to a sink. Apply the hand sanitizer as you would hand lotion, making sure to cover the fronts and backs of both hands. Use a product that contains at least 60 percent alcohol.

Gastroenteritis

When you have diarrhea and vomiting, you may say you have the "stomach flu." These symptoms often are due to a condition called gastroenteritis.

With gastroenteritis, your stomach and intestines are irritated and inflamed. The cause is typically a viral or bacterial infection.

Symptoms of Gastroenteritis

With gastroenteritis, the main symptoms you probably have are watery diarrhea and vomiting. You might also have stomach pain, cramping, fever, nausea, and a headache.

Because of diarrhea and vomiting, you also can become dehydrated. Watch for signs of dehydration, such as dry skin and a dry mouth, feeling lightheaded, and being really thirsty. Call your doctor if you have any of these symptoms.

Stomach Flu and Children

Children can get dehydrated quickly, so if your child has the stomach flu, it's important that you look for signs that he is very thirsty or has dry skin or a dry mouth. If you have a baby, look for fewer, drier diapers.

Keep children with gastroenteritis out of day care or school until all symptoms are gone. Check with your doctor before giving your child any medicine. Drugs used to control diarrhea and vomiting aren't usually given to children younger than 5.

To help prevent rotavirus the most common cause of stomach flu for children there are two vaccines that can be given to infants.

Causes of Gastroenteritis

There are many ways gastroenteritis can be spread:

- Contact with someone who has the virus.
- Contaminated food or water.
- Unwashed hands after going to the bathroom or changing a diaper.

The most common cause of gastroenteritis is a virus. Gastroenteritis flu can be caused by many different kinds of viruses. The main types are rotavirus and norovirus.

Rotavirus is the world's most common cause of diarrhea in infants and young children. Norovirus is the most common cause of serious gastroenteritis and also foodborne disease outbreaks in the U.S.

Although not as common, bacteria such as E. coli and salmonella can also trigger the stomach flu. Salmonella and campylobacter bacteria are the most common bacterial causes of gastroenteritis in the U.S. and are usually spread by undercooked poultry, eggs, or poultry juices. Salmonella can also be spread through pet reptiles or live poultry.

Cholera

Cholera is a bacterial disease usually spread through contaminated water. Cholera causes severe diarrhea and dehydration. Left untreated, cholera can be fatal in a matter of hours, even in previously healthy people.

Modern sewage and water treatment have virtually eliminated cholera in industrialized countries. The last major outbreak in the United States occurred in 1911. But cholera is still present in Africa, Southeast Asia and Haiti. The risk of cholera epidemic is highest when poverty, war or natural disasters force people to live in crowded conditions without adequate sanitation.

Cholera is easily treated. Death results from severe dehydration that can be prevented with a simple and inexpensive rehydration solution.

Symptoms

Most people exposed to the cholera bacterium (Vibrio cholerae) don't become ill and never know they've been infected. Yet because they shed cholera bacteria in their stool for seven to 14 days, they can still infect others through contaminated water. Most symptomatic cases of cholera cause mild or moderate diarrhea that's often hard to distinguish from diarrhea caused by other problems.

Only about 1 in 10 infected people develops more-serious signs and symptoms of cholera, usually within a few days of infection.

Symptoms of cholera infection may include:

- Diarrhea: Cholera-related diarrhea comes on suddenly and may quickly cause dangerous fluid loss — as much as a quart (about 1 liter) an hour. Diarrhea due to cholera often has a pale, milky appearance that resembles water in which rice has been rinsed (rice-water stool).
- Nausea and vomiting: Occurring especially in the early stages of cholera, vomiting may persist for hours at a time.
- Dehydration: Dehydration can develop within hours after the onset of cholera symptoms. Depending on how many body fluids have been lost, dehydration can range from mild to severe. A loss of 10 percent or more of total body weight indicates severe dehydration.

  Signs and symptoms of cholera dehydration include irritability, lethargy, sunken eyes, a dry mouth, extreme thirst, dry and shriveled skin that's slow to bounce back when pinched into a fold, little or no urine output, low blood pressure, and an irregular heartbeat (arrhythmia).

Dehydration may lead to a rapid loss of minerals in your blood (electrolytes) that maintain the balance of fluids in your body. This is called an electrolyte imbalance.

Electrolyte Imbalance

An electrolyte imbalance can lead to serious signs and symptoms such as:

- Muscle cramps: These result from the rapid loss of salts such as sodium, chloride and potassium.

- Shock: This is one of the most serious complications of dehydration. It occurs when low blood volume causes a drop in blood pressure and a drop in the amount of oxygen in your body. If untreated, severe hypovolemic shock can cause death in a matter of minutes.

Signs and Symptoms of Cholera in Children

In general, children with cholera have the same signs and symptoms adults do, but they are particularly susceptible to low blood sugar (hypoglycemia) due to fluid loss, which may cause:

- An altered state of consciousness.
- Seizures.
- Coma.

Causes

A bacterium called Vibrio cholerae causes cholera infection. However, the deadly effects of the disease are the result of a potent toxin called CTX that the bacterium produce in the small intestine. CTX binds to the intestinal walls, where it interferes with the normal flow of sodium and chloride. This causes the body to secrete enormous amounts of water, leading to diarrhea and a rapid loss of fluids and salts (electrolytes).

Contaminated water supplies are the main source of cholera infection, although raw shellfish, uncooked fruits and vegetables, and other foods also can harbor V. cholerae.

Cholera bacteria have two distinct life cycles one in the environment and one in humans.

Cholera Bacteria in the Environment

Cholera bacteria occur naturally in coastal waters, where they attach to tiny crustaceans called copepods. The cholera bacteria travel with their hosts, spreading worldwide as the crustaceans follow their food source certain types of algae and plankton that grow explosively when water temperatures rise. Algae growth is further fueled by the urea found in sewage and in agricultural runoff.

Cholera Bacteria in People

When humans ingest cholera bacteria, they may not become sick themselves, but they still pass the bacteria in their stool. When human feces contaminate food and water supplies, both can serve as ideal breeding grounds for the cholera bacteria.

Because more than a million cholera bacteria approximately the amount you'd find in a glass of contaminated water are needed to cause illness, cholera usually isn't transmitted through casual person-to-person contact.

The most common sources of cholera infection are standing water and certain types of food, including seafood, raw fruits and vegetables, and grains.

Surface or well water: Cholera bacteria can lie dormant in water for long periods, and contaminated public wells are frequent sources of large-scale cholera outbreaks. People living in crowded conditions without adequate sanitation are especially at risk of cholera.

- Seafood: Eating raw or undercooked seafood, especially shellfish, that originates from certain locations can expose you to cholera bacteria. Most recent cases of cholera occurring in the United States have been traced to seafood from the Gulf of Mexico.
- Raw fruits and vegetables: Raw, unpeeled fruits and vegetables are a frequent source of cholera infection in areas where cholera is endemic. In developing nations, uncomposted manure fertilizers or irrigation water containing raw sewage can contaminate produce in the field.
- Grains: In regions where cholera is widespread, grains such as rice and millet that are contaminated after cooking and allowed to remain at room temperature for several hours become a medium for the growth of cholera bacteria.

Risk Factors

Everyone is susceptible to cholera, with the exception of infants who derive immunity from nursing mothers who have previously had cholera. Still, certain factors can make you more vulnerable to the disease or more likely to experience severe signs and symptoms. Risk factors for cholera include:

- Poor sanitary conditions: Cholera is more likely to flourish in situations where a sanitary environment including a safe water supply — is difficult to maintain. Such conditions are common to refugee camps, impoverished countries, and areas devastated by famine, war or natural disasters.
- Reduced or nonexistent stomach acid (hypochlorhydria or achlorhydria): Cholera bacteria can't survive in an acidic environment, and ordinary stomach acid often serves as a first line defense against infection. But people with low levels of stomach acid such as children, older adults, and people who take antacids, H-2 blockers or proton pump inhibitors lack this protection, so they're at greater risk of cholera.
- Household exposure: You're at significantly increased risk of cholera if you live with someone who has the disease.
- Type O blood: For reasons that aren't entirely clear, people with type O blood are twice as likely to develop cholera compared with people with other blood types.
- Raw or undercooked shellfish: Although large-scale cholera outbreaks no longer

occur in industrialized nations, eating shellfish from waters known to harbor the bacteria greatly increases your risk.

Complications

Cholera can quickly become fatal. In the most severe cases, the rapid loss of large amounts of fluids and electrolytes can lead to death within two to three hours. In less extreme situations, people who don't receive treatment may die of dehydration and shock hours to days after cholera symptoms first appear.

Although shock and severe dehydration are the most devastating complications of cholera, other problems can occur, such as:

- Low blood sugar (hypoglycemia): Dangerously low levels of blood sugar (glucose) — the body's main energy source may occur when people become too ill to eat. Children are at greatest risk of this complication, which can cause seizures, unconsciousness and even death.
- Low potassium levels (hypokalemia): People with cholera lose large quantities of minerals, including potassium, in their stools. Very low potassium levels interfere with heart and nerve function and are life-threatening.
- Kidney (renal) failure: When the kidneys lose their filtering ability, excess amounts of fluids, some electrolytes and wastes build up in your body — a potentially life-threatening condition. In people with cholera, kidney failure often accompanies shock.

Prevention

Cholera is rare in the United States with the few cases related to travel outside the U.S. or to contaminated and improperly cooked seafood from the Gulf Coast waters.

If you're traveling to cholera-endemic areas, your risk of contracting the disease is extremely low if you follow these precautions:

- Wash hands with soap and water frequently, especially after using the toilet and before handling food. Rub soapy, wet hands together for at least 15 seconds before rinsing. If soap and water aren't available, use an alcohol-based hand sanitizer.
- Drink only safe water, including bottled water or water you've boiled or disinfected yourself. Use bottled water even to brush your teeth. Hot beverages are generally safe, as are canned or bottled drinks, but wipe the outside before you open them. Avoid adding ice to your beverages unless you made it yourself using safe water.
- Eat food that's completely cooked and hot and avoid street vendor food, if

possible. If you do buy a meal from a street vendor, make sure it's cooked in your presence and served hot.

- Avoid sushi, as well as raw or improperly cooked fish and seafood of any kind.
- Stick to fruits and vegetables that you can peel yourself, such as bananas, oranges and avocados. Stay away from salads and fruits that can't be peeled, such as grapes and berries.
- Be wary of dairy foods, including ice cream, which is often contaminated, and unpasteurized milk.

Typhoid

Typhoid fever also called typhoid is an acute infectious disease caused by the bacterium Salmonella enterica serovar Typhi. The bacterium usually enters the body through the mouth by the ingestion of contaminated food or water, penetrates the intestinal wall, and multiplies in lymphoid tissue; it then enters the bloodstream and causes bacteremia.

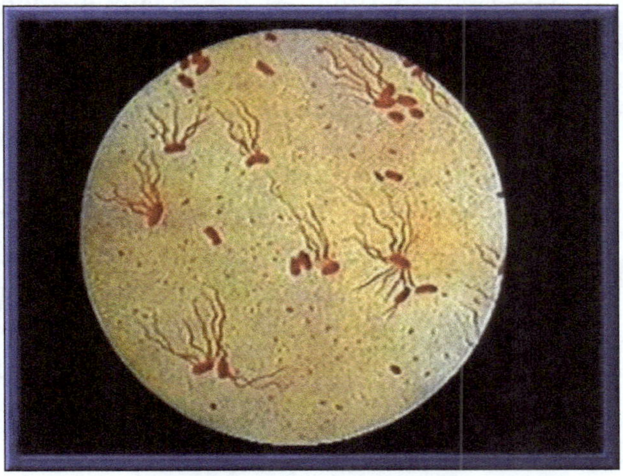

Salmonella typhi: Photomicrograph of salmonella typhi, the causative agent of typhoid fever. Centers for Disease Control and Prevention (CDC).

Most major epidemics of typhoid fever have been caused by the pollution of public water supplies. Food and milk may be contaminated, however, by a human carrier of the disease who is employed in handling and processing them; by flies; or by the use of polluted water for cleaning purposes. Shellfish, particularly oysters, grown in polluted water and fresh vegetables grown on soil fertilized or contaminated by untreated sewage are other possible causes.

The prevention of typhoid fever depends mainly on proper sewage treatment, filtration and chlorination of water, and exclusion of carriers from employment in food industries and restaurants. In the early part of the 20th century, prophylactic vaccination using killed typhoid organisms was introduced, mainly in military forces and institutions, and contributed to a lowering of the incidence of the disease.

Course of Infection

After an average 10–14-day incubation period, the early symptoms of typhoid appear: Headache, malaise, generalized aching, fever, and restlessness that may interfere with sleep. There may be loss of appetite, nosebleeds, cough, and diarrhea or constipation. Persistent fever develops and gradually rises, usually in a stepwise fashion, reaching a peak of 39 or 40 °C (103 or 104 °F) after 7–10 days; left untreated, the fever continues with only slight morning remissions for another 10–14 days, sometimes longer.

During about the second week of fever, typhoid bacilli are present in great numbers in the bloodstream. At that point, some patients develop a rash of small rose-coloured spots on the trunk, which lasts four or five days and then fades away. The lymph follicles (Peyer patches) along the intestinal wall in which the typhoid bacilli have multiplied become inflamed and necrotic and may slough off, leaving ulcers in the walls of the intestine. The dead fragments of intestinal tissue may erode blood vessels, causing hemorrhage, or they may perforate the intestinal wall, allowing the intestine's contents to enter the peritoneal cavity (peritonitis). Other complications can include acute inflammation of the gallbladder, heart failure, pneumonia, osteomyelitis, encephalitis, and meningitis. With a continued high fever, the symptoms usually increase in intensity, and mental confusion and delirium may appear.

By the end of the third week, the patient is emaciated, abdominal symptoms are marked, and mental disturbance is prominent. In favourable cases, about the beginning of the fourth week, the fever begins to decline, the symptoms begin to abate, and the temperature gradually returns to normal. If untreated, typhoid fever proves fatal in about 10 to 30 percent of all cases; with treatment, as few as 1 percent of patients die from the disease. Patients with diseases such as cancer or sickle cell anemia are particularly prone to develop serious and prolonged infection with S. Typhi.

Carriers

Typhoid bacteria can persist for an indefinite period of time in the bile passages of patients. If they practice poor hygiene or if they are food handlers, those carriers can pass the infection to healthy persons. Patients who are recovering from typhoid fever are transient carriers of the disease, excreting the causative bacteria in the stool or urine for up to three months. Patients who continue to excrete the bacteria for a year or more after infection are considered to be long-term carriers; those individuals harbour the microorganisms and typically shed them for years.

One of the most famous instances of carrier-borne disease in medical history was the case of "Typhoid Mary" (byname of Mary Mallon). Fifty-one original cases of typhoid and three deaths were directly attributed to her during the early 20th century.

References

- Food-allergens: allergenbureau.net, Retrieved 09 March 2018
- Food-safety-and-types-food-contamination: foodsafety.ca, Retrieved 16 July 2018
- Physical-contaminants, services: campdenbri.co.uk, Retrieved 09 August 2018
- Amebiasis, health: healthline.com, Retrieved 31 August 2018
- Typhoid-fever, science: britannica.com, Retrieved 11 March 2018

Permissions

All chapters in this book are published with permission under the Creative Commons Attribution Share Alike License or equivalent. Every chapter published in this book has been scrutinized by our experts. Their significance has been extensively debated. The topics covered herein carry significant information for a comprehensive understanding. They may even be implemented as practical applications or may be referred to as a beginning point for further studies.

We would like to thank the editorial team for lending their expertise to make the book truly unique. They have played a crucial role in the development of this book. Without their invaluable contributions this book wouldn't have been possible. They have made vital efforts to compile up to date information on the varied aspects of this subject to make this book a valuable addition to the collection of many professionals and students.

This book was conceptualized with the vision of imparting up-to-date and integrated information in this field. To ensure the same, a matchless editorial board was set up. Every individual on the board went through rigorous rounds of assessment to prove their worth. After which they invested a large part of their time researching and compiling the most relevant data for our readers.

The editorial board has been involved in producing this book since its inception. They have spent rigorous hours researching and exploring the diverse topics which have resulted in the successful publishing of this book. They have passed on their knowledge of decades through this book. To expedite this challenging task, the publisher supported the team at every step. A small team of assistant editors was also appointed to further simplify the editing procedure and attain best results for the readers.

Apart from the editorial board, the designing team has also invested a significant amount of their time in understanding the subject and creating the most relevant covers. They scrutinized every image to scout for the most suitable representation of the subject and create an appropriate cover for the book.

The publishing team has been an ardent support to the editorial, designing and production team. Their endless efforts to recruit the best for this project, has resulted in the accomplishment of this book. They are a veteran in the field of academics and their pool of knowledge is as vast as their experience in printing. Their expertise and guidance has proved useful at every step. Their uncompromising quality standards have made this book an exceptional effort. Their encouragement from time to time has been an inspiration for everyone.

The publisher and the editorial board hope that this book will prove to be a valuable piece of knowledge for students, practitioners and scholars across the globe.

Index

www.ingramcontent.com/pod-product-compliance
Lightning Source LLC
LaVergne TN
LVHW080634210926
848485LV00003B/262